Springer Series in Reliability Engineering

Series Editor

Hoang Pham, Department of Industrial and Systems Engineering,
Rutgers University, Piscataway, NJ, USA

More information about this series at https://link.springer.com/bookseries/6917

Tarannom Parhizkar · Ingrid B. Utne ·
Jan-Erik Vinnem

Online Probabilistic Risk Assessment of Complex Marine Systems

Principles, Modelling and Applications

Springer

Tarannom Parhizkar
Department of Marine Technology
Norwegian University of Science
and Technology
Trondheim, Norway

B. John Garrick Institute for
the Risk Sciences
University of California
Los Angeles, CA, USA

Jan-Erik Vinnem
Department of Marine Technology
Norwegian University of Science
and Technology
Trondheim, Norway

Ingrid B. Utne
Department of Marine Technology
Norwegian University of Science
and Technology
Trondheim, Norway

ISSN 1614-7839 ISSN 2196-999X (electronic)
Springer Series in Reliability Engineering
ISBN 978-3-030-88097-2 ISBN 978-3-030-88098-9 (eBook)
https://doi.org/10.1007/978-3-030-88098-9

This Springer imprint is published by the registered company Springer Nature Switzerland AG
The registered company address is: Gewerbestrasse 11, 6330 Cham, Switzerland

Preface

This book provides a general framework for online risk assessment considering human, software, and hardware interactions in automated and autonomous complex systems, with particular focus on marine applications. There is a significant interest in industry, authorities, and academia on the prospects of complex autonomous systems. Much work focuses on technological developments and discussions on economic, legal, and safety implications. There has been less focus so far on risk assessment and modeling concerning the conceptualization, design, and, in particular on the safe operation of these systems.

The safe operation of automated and autonomous complex systems requires close coordination between human operators, organization, and the technical systems and components. Often, both the systems and human operators face dynamic decision-making conditions with incomplete information and little time to consider options. Sophisticated techniques and methodologies have been developed to analyze different complex systems, and advances have been made in fundamental issues. Most of the current techniques for risk assessment, however, largely rely on a static analysis of a complex system. Unfortunately, static analyses cannot capture the dynamic factors and feedback loops that influence the system behavior of automated and autonomous systems.

In recent years, simulation-based approaches have been proposed that capture dynamic interactions and couple the human, hardware, and software performance to provide a better prediction of hazardous events and evaluation of risks related to operator(s) and system behavior. So far, such methods have been mostly utilized for nuclear power plants and for space missions. There is a growing need for dynamic and online risk assessment tools capable of providing information about risk for decision support of automated and autonomous systems, for example, in the marine domain. This demand is due to the increase of autonomous functionality that is being incorporated, which makes systems more complex and risk assessment more challenging. Furthermore, trust and approval by authorities and the public require safe design and operation of the autonomous systems.

The main intended audiences of the book are marine, maritime and offshore petroleum industry professionals, aerospace industry professionals, nuclear power industry, consultants, authorities, researchers, and graduate students in the field of engineering, in particular for those interested in risk and reliability assessment of complex systems. The framework of this book could be used in the design of decision support tools to help operators make better decisions; for development of supervisory risk control and improved intelligence of autonomous systems, and generally, to improve complex systems' safety.

The content of the book has been researched and developed by the authors over several years in different research and industry projects. A major leap forward, however, was achieved from 2018 to 2020 when Dr. Parhizkar was a postdoc at the Norwegian University of Science and Technology (NTNU).

Trondheim, Norway

Tarannom Parhizkar
Ingrid B. Utne
Jan-Erik Vinnem

Contents

Abbreviations

ADAPT	Analysis of dynamic accident progression trees
ADS	Accident dynamic simulator
AHTS	Anchor handling
AI	Artificial intelligence
AROVs	Autonomous remotely operated vehicles
ATHEANA	A technique for human event analysis
BN	Bayesian network
BVP	Blood volume pulse
CAISO	California Independent System Operator
CAMEO	Cognitive and action modeling of an erring operator
CES	Cognitive environment simulation
COSIMO	Cognitive simulation model
CREAM	Cognitive reliability error analysis method
CS	Complex system
CSRM	Context-based software risk assessment Methodology
DBN	Dynamic Bayesian network
DC	Disconnection
DESDs	Dynamic event sequence diagrams
DFM	Dynamic flow graph methodology
DGPS	Differential Global Positioning System
DP	Dynamic positioning
DPO	Dynamic positioning operators
DPRA	Dynamic probabilistic risk assessment
DSV	Diving support vessels
DYLAM	Dynamical logical analytical methodology
ESD(s)	Event sequence diagram(s)
ESD	Emergency shutdown
FMEA	Failure mode and effect analysis
FPSOs	Floating production storage and offloading
FSL	Flow state logic
FSUs	Floating storage units

FTA	Fault tree analysis
FTs	Fault trees
GIG	Gaussian distribution
HCL	Hybrid causal logic modeling
HEP	Human error probability
HMI	Human–machine interface
HR	Heart rate
HRA	Human reliability assessment
HRM	High-reliability management
IDA	Information, decision, and action
IDAC	Information, decision, and action crew
IMCA	International Marine Contractors Association
IMO	International Maritime Organizations
IMR	Inspection, maintenance, and repair
INTEROPS	Integrated reactor operator/system
LBL	Long baseline
LoA	Levels of autonomy
LOP	Loss of position
MASS	Maritime autonomous surface ships
MD	Mechanical damage
MODU	Mobile offshore drilling unit
MSWB	Main switchboard
MV	Medium voltage
NASA	National Aeronautics and Space Administration
NIST	National Institute of Standards and Technology
OCV	Offshore construction vessels
OIM	Offshore installation managers
PCB	Printed circuit board
PLC	Programmable logic controller
PSA, PRA	Probabilistic safety/risk analysis
PSFs	Performance shaping factors
PSO	Particle swarm optimization
PSV	Platform supply vessels
QRA	Quantitative risk assessment
RAVEN	Risk analysis virtual environment
ROVs	Remotely operated vehicles
RPD	Recognition primed naturalistic decision-making
SA	Situation awareness
SEAMAID	Simulation-based evaluation and analysis support system for man–machine interface design
SJA	Safe job analysis
SPAR-H	Standardized plant analysis risk-human reliability analysis
SRA	Society for Risk Analysis
ST	Shuttle tankers
STAMP	Systems theoretic accident model and processes

STPA	Systems theoretic process analysis
SYBORG	Simulation system for behavior of an operating group
WSOG	Well-specific operating guidelines

Chapter 1
Introduction

1.1 Problem Statement

There is a significant interest in industry, authorities, and academia on the prospects of marine and maritime autonomous systems. Much work focuses on technological developments and discussions on economic, legal, and safety implications [1]. There has been less focus so far on risk assessment and modeling concerning the conceptualization, design, and, in particular; on the safe operation of these systems.

The safe operation of automated and autonomous marine and maritime systems requires close coordination between human operators, (onshore) organization, and the technical systems and components. For example, dynamic positioning operators (DPOs) must efficiently perform complex tasks following an incident to maintain the position of a vessel and provide an adequate level of safety. Often, operators of complex systems face dynamic decision-making conditions with incomplete information and little time to consider options. Sophisticated techniques and methodologies have been developed to analyze different complex systems, and advances have been made in fundamental issues, as presented in [2, 3]. Most of the current techniques for risk assessment, however, largely rely on a static analysis of a complex system. Unfortunately, static analyses cannot capture the dynamic factors and feedback loops that influence the system behavior of automated and autonomous systems [4–10].

In recent years, simulation-based approaches have been proposed that capture the dynamic interactions and couple the human and hardware performance to provide a better analysis and estimation of the risks related to operator and system behavior. So far, such methods have been mostly utilized in nuclear power plants and for space missions.

There is a growing need for online risk assessment tools capable of providing information about risk for decision support of autonomous systems [12, 12], such as in the marine and maritime industry. This demand is due to the increase of autonomous functions that are being incorporated, which makes systems more complex and risk

T. Parhizkar et al., *Online Probabilistic Risk Assessment of Complex Marine Systems*, Springer Series in Reliability Engineering,
https://doi.org/10.1007/978-3-030-88098-9_1

assessment more challenging. Trust and approval by authorities and the general public require safe design and operation of the autonomous systems.

In this book, a general framework for online risk assessment considering human, software, and hardware interactions in automated and autonomous marine and maritime systems is presented. The framework is applied to multiple case studies, and the usefulness is discussed. The framework should be applicable to a lot of different application areas, beyond the marine and maritime domain.

1.2 Background

In recent years, multiple studies have analyzed the safety and risk management challenges of autonomous marine and maritime systems. Some of these studies are related to the risks associated with changes in manning and organizational modifications. In [13, 14], related legal challenges are presented and discussed. Chang et al. [14] presents and discusses the legal status of the unmanned maritime vehicles at the time. In [15–17], the challenges for transferring the role of operators involved in the vessel to the operational context of unmanned vessels are discussed. Hegde et al. [18] presents a process utilized to develop safety envelopes and subsea traffic rules for autonomous remotely operated vehicles (AROVs) used in subsea inspection, maintenance, and repair (IMR) operations.

Thieme et al. [19] investigates to which extent the current vessel risk models for ship–structure collision, ship–ship collision, and groundings are applicable for risk assessment of Marine Autonomous Surface Ships (MASS). They found that few risk assessment models are not directly applicable as they lack the consideration of the communication connection with the shore control center, impacts from software failures on system risk, and interaction between conventional and autonomous systems. In [20], the feasibility of using a systems approach to analyze and evaluate the risk level in dynamically positioned systems are investigated. An adapted version of the system-theoretic process analysis is presented and applied to an example. This study demonstrates that the reliability-centered approaches, such as the failure mode and effect analysis (FMEA), sea trials, and hardware-in-the loop testing, are insufficient to ensure safety. For example, to perform a comprehensive safety assessment of dynamically positioned (DP) systems, multiple safety constraints in addition to component failures should be considered.

In [21], a process for evaluating hazard analysis and management that provides safety input information for the design of autonomous vessels is proposed. Vinnem and Utne [22] emphasize that autonomous ships will be exposed to the threat of cyberattacks, but most critical hazards might possibly not be associated with consequences for the ship itself or its cargo but be a threat to critical infrastructure along the coast and offshore—if a ship under alien command is used as a “battering ram” to cause major structural damage. A study in [23] presents a risk framework named MaCRA used to analyze cyber-risk of autonomous ships.

Zhang et al. [24] proposes a fuzzy Bayesian network to quantify the operational risk of unmanned vessels in inland rivers. The proposed Bayesian network considers different operational and environmental factors that affect vessel operations. The output of the model could provide early warning to take appropriate preventive measures to improve the overall safety of operation. In [25], a Bayesian network is proposed to evaluate the probability of aborting an autonomous subsea inspection, maintenance, and repair (IMR) operation. The proposed Bayesian network considers technical, organizational, and operational interactions with different mission scenarios. Results show that the proposed model may provide a useful tool to human supervisors in their decision-making processes. Hegde et al. [18] proposes risk-based indicators to assist decision support systems in autonomous remotely operated vehicles (AROVs), including time to collision, mean time to collision, and mean impact energy.

Hegde et al. [26] discusses the importance of considering human and organizational factors in risk assessment of the offshore oil and gas industry. In [27], the importance of human factors is highlighted, and a method is proposed to consider human-autonomous system collaboration in system risk assessment. In [28], a dynamic risk model for dynamic positioning (DP) systems is presented that shows the effects of considering operators' behavior, in addition to component failures, on system overall risk level. The proposed model is applied to a DP system as an example, and results are discussed. This study illustrates the importance of considering dynamic behavior of operators and components' operation in system risk evaluation. Parhizkar et al. [29] analyzes the main real-time risks in operation of autonomous marine systems at various level of autonomy. In this research, the high-reliability management (HRM) framework is applied to two case studies with different operational states that vary across different level of autonomy, each state of operations having significantly different risks to be managed in real time.

Most of the reviewed literature focus on a static environment, i.e., the impact of the dynamic behavior of operators, environmental conditions, system operation, and their interactions over time are simplified or not considered. However, different studies have shown that dynamic behavior of a complex system affects the system risk level significantly [30–32].

The reviewed literature shows the need for developing online risk management of automated and autonomous marine and maritime systems with different levels of autonomy that considers system, environmental, and operational interactions. Hence, in this book, a general framework for online risk assessment of automated and autonomous marine systems is presented. This framework considers hardware, human, and software dynamic interactions. The framework aims to help designers and provide operators with health monitoring and risk assessment for marine and maritime systems with different levels of autonomy.

1.3 Safety of Autonomous Marine and Maritime Systems

The control functionality of many manned systems, such as offshore oil and gas rigs and ships with complex automation and dynamic positioning (DP) systems, may be characterized as autonomous. Failures of DP systems may cause major accidents. Therefore, many examples in this book are linked to DP systems and operations.

Often, autonomous systems and operations are associated with unmanned systems, but both manned and unmanned systems can operate in a low to high autonomy level, i.e., from manual and automatic to "intelligently" and independently. Hence, it is useful to distinguish between unmanned systems and autonomy. The National Institute of Standards and Technology (NIST) [33] defines unmanned systems as a powered physical system, with no human operator aboard the principal components. It acts in the physical world for achieving assigned tasks. It may be mobile or stationary and can include all associated supporting components. Autonomy, on the other hand, can be defined as a system's or subsystem's own ability of integrated sensing, perceiving, analyzing, communicating, planning, decision-making, and acting, to achieve its goals as assigned by its human operator(s) through designed human–machine interface (HMI). This definition is based on NIST [33] but adjusted for autonomous systems and operations, both manned and unmanned. This means that unmanned systems are a "subcategory" of autonomous systems.

Levels of autonomy (LoA) (Table 1.1) can be defined as a set of metrics that describe detailed aspects of an autonomous system and operation, including operator dependency, communication structure, human–machine interface (HMI), an online risk management system, intelligence, planning functionalities, and mission complexity [34]. For complex systems, such as ships, the advancement is not from fully manned and manual operation to fully unmanned and highly autonomous operation. Rather there is a gradual transition toward higher LoA. In an operational environment with a combination of manned and unmanned autonomous systems, safe operations may be challenging since the behavior of a human operator represents an uncertainty that the unmanned and autonomous system need to cope with and vice versa. Hence, the LoA impacts system and operational risk. In this book, we use the term automated for systems with low LoA and autonomous for systems with higher LoA (level 2-4 in Table 1.1).

Many systems today have functionality in between low and high autonomy, such as advanced power and propulsion systems, including DP, but the technology is moving toward higher levels of autonomy for an increasing number of applications [34]. An increased level of automation and autonomy may improve safety, efficiency, and performance in operation, supporting the human operator in decision-making and supervision and reducing human workload. DP systems enable several offshore oil and gas operations. A motivation for autonomous ships is reduced building and operational costs as the ships can be redesigned and operations can be more efficient.

Safety is usually taken to imply protection of human life and health, the environment and sometimes also assets, where protection of life and health is the primary concern. There would be no human life to protect a highly autonomous system

Table 1.1 Levels of autonomy and overall risk aspects (Based on [34])

Autonomy level	Description	Overall risk aspects
1: Automatic operation (Remote control)	The system operates automatically. The human operator directs and controls all high-level mission planning functions, often preprogrammed. System states, environmental conditions, and sensor data are presented to the operator through a human–machine interface (HMI) (human-in-the-loop/human operated)	Operator experience, procedures, and training are essential. The risk is highly related to hardware and software errors and design flaws, operator performance, workload, external conditions, and the probability of the operator making an erroneous decision. HMI is important to ensure situation awareness, including robust high bandwidth communication link between the system and human operator. The associated control system software at any level of the system also represents a risk
2: Management by consent	The system automatically makes recommendations for mission or process actions related to specific functions, and the system prompts the human operator at important points in time for information or decisions. At this level, the system may have limited communication bandwidth, including time delay, due to, e.g., distance. The system can perform many functions independently of human control when delegated to do so (human-delegated)	In addition to the risk mentioned above, the importance and impact of HMI, as well as a robust communication link, increasingly impact the probability of an erroneous decision. Software and the control system software, including sensors, constitute an increasing risk
3: Semi-autonomous operation or management by exception	The system automatically executes mission-related functions when response times are too short for human intervention. The human may override or change parameters and cancel or redirect actions within defined timelines. The operator's attention is only brought to exceptions for certain decisions (human-supervisory control)	Risk is highly dependent on the situation awareness capabilities of the system, as well as the operator. The risk related to an erroneous operator action may be lower, but a challenge is related to unforeseen incidents and consequences the system is not designed for and not capable of anticipating, as well as boredom for the human operator. Mission complexity and environmental complexity may increase. Complexity challenges the verification of safe performance of the system. The control system software creates a significant risk

(continued)

Table 1.1 (continued)

Autonomy level	Description	Overall risk aspects
4: Highly autonomous operation	The system automatically executes mission or process related functions in an unstructured environment with ability to plan and replan the mission or process. The human may be informed about the progress. The system is independent and "intelligent."	Risk reduction is dependent on a robust and resilient design of the system, but also on an online risk management system. Robust support systems (navigation, control, etc.,) are needed and situation awareness for the system is related to its perception of the situation. Hence, risk must be considered for both internal and external conditions of the system. Increased system complexity, including the control system software, is a major obstacle for verification of safe performance of the system. Efficient and high integrity and adaptive functionality are very important

without any manning present. This is not likely to be the case in the foreseeable future, manning would at least be present for some periods. DP vessels also have crew onboard. In practice, there is always human life and health to be protected. Safety is thus considered in the traditional context in this book.

Risk assessment usually involves two dimensions; the uncertainty about future outcomes, normally expressed with probabilities, as well as consequences of future outcomes, analyzed by special consequence tools. This book does not address consequences, as they are in any case assumed to be severe in case of collision or other types of major marine hazards. This book is therefore limited to the probabilities of future outcomes.

Advanced control systems with autonomy functions, such as DP, may cause complexity and interlocks that are hard to identify and analyze. New types of risks may be introduced, due to unexpected interdependencies in the system design, operation complexity, and environmental challenges. Shutting down and remobilizing the operation of an advanced system, such as a rig or ship, due to hazards caused by automatic/autonomous systems or false alarms, are not viable. Systems with embedded software and functional integration are complex, because physical separation and segregation of components, such as redundancy in ship machinery systems, may be overruled by software and control systems that operate across physical boundaries.

Risk assessment of autonomous systems is not a simple task, because it needs to capture the dynamic and complex aspects of such a variety of systems, as well as uncertainties related to lack of knowledge from operational experience. Risk assessments should be carried out both during the design and operation of a system, as shown in Fig. 1.1. For autonomous systems with embedded software solutions, upgrades may take place for bug fixes and improved functionalities. This should

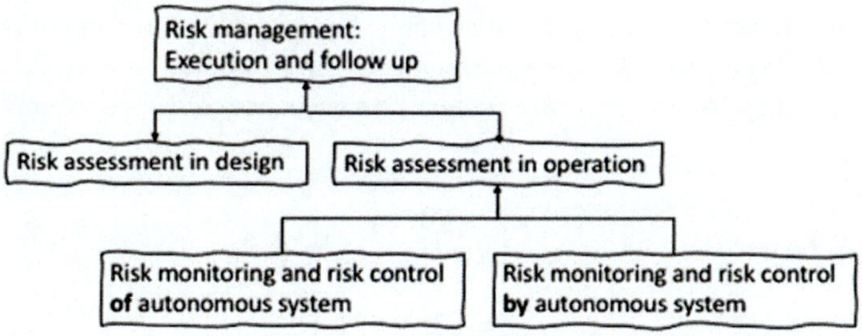

Fig. 1.1 Risk assessment and risk control of autonomous systems and operations (Based on [34])

also initiate testing and verification before the system is put into operation again to reveal any complex dependencies and malfunctions. During operation, risk can be assessed and controlled by the operator and/or by the autonomous system; the extent determined by the system's level of autonomy. Hence, risk control related to autonomous systems can occur in two equally important and complementary ways, depending on a system's autonomy level: (i) risk control of the autonomous system performed by the human operator(s), which is mostly relevant for low LoA; (ii) risk control performed by the autonomous system itself, which is increasingly important in the higher LoA, also called supervisory risk control, [35].

1.4 Dynamic Positioning Systems and Safety

1.4.1 Background

The first automatic course steering device for ships was constructed in 1911. It was able to adjust for various sea states through feedback control and automatic gain adjustments. From then on, there has been a technological evolution, in particular after the emergence of DP vessels in the 1960s. DP systems were developed to enable offshore oil and gas drilling in deeper waters where it was not possible to use anchor lines to fix the position of rigs and vessels.

Dynamic positioning (DP) vessels are defined by the International Maritime Organizations (IMO) as "a unit or a vessel which automatically maintains its position and/or heading (fixed location, relative location or predetermined track) by means of thruster force".

By 1975, there were twenty DP vessels, of which ten were offshore drilling vessels, which had increased to 65 in 1980. Nowadays, there are more than 3000 DP systems in operation worldwide, which is a core technology in oil and gas offshore operations. Examples of vessels with DP are mobile offshore drilling units (MODU), shuttle tankers (ST), offshore construction vessels (OCV), flotels, intervention vessels (IMR), diving support vessels (DSV), anchor handling (AHTS), survey, dredging and rock dumping, pipe- and cable laying, and platform supply vessels (PSV). In addition, DP (including automatic sailing systems) are utilized in other industries, such as for cruise ships, nuclear power plants, and space shuttles.

1.4.2 Safety

An essential safety factor for autonomous marine and maritime systems is maintaining its position, often obtained through a dynamic positioning (DP) system. A DP vessel automatically keeps its position in surge and sway and its heading (yaw) exclusively by means of active thrusters and propellers. A DP system includes a power system, a thruster system, and DP control system. The dynamic positioning operator (DPO) can be considered an integral part of the DP system [35].

The power system of a DP vessel usually consists of diesel engines or gas turbines that are connected to generators producing electrical power that is distributed to consumer units, e.g., thrusters. The thruster system provides forces and the desired motion of the vessel, coordinated and controlled by the DP control system. The DP control system determines the suitable thrust force in surge, sway, and yaw, based on feedback control, and allocates thrust responsibility to each thruster. The DPO configures the power and control systems by selecting set points and operating modes. For more details on the DP system, see, e.g., Utne et al. [35].

Loss of position is the main safety concern with DP systems. Already in 1970, Honeywell included computer redundancy onboard the drilling vessel SEDCO, using one active and one passive computer. The purpose was to provide a minimum of 150 h continuous operation. In 1987, the first triple-redundant DP system was launched [35, 36].

Safety of DP systems is traditionally regulated through stipulating redundancy requirements for the technical systems. The concept of technical redundancy has been common in reliability engineering since World War I, and the maritime industry uses technical redundancy as a means of controlling risk, for example, IMO's guidelines for DP systems. Breivik et al. [37] has defined three DP classes, with the basis in the worst single failure modes.

- Equipment Class 1 (DP1): Loss of position may occur after a single failure,
- Equipment Class 2 (DP2): Loss of position should not occur after a single failure of an active component,
- Equipment Class 3 (DP3): Loss of position should not occur from any single failure (including flooded watertight component and fire in subdivision).

These DP classes and requirements are also implemented by classification societies. Sufficient number of generators, thrusters, and reference systems shall be functional in order to fulfill the redundancy requirements defined by the DP class, so that the failure of one component does not result in a loss of position.

The equipment class is usually decided based on industry practice or the relevant coastal state and administration. The IMO guidelines [37] state that the equipment class for an operation should be agreed upon by the vessel owner and the customer with regard to the potential consequences of loss of position.

A consequence analysis should be incorporated in the system for DP Class 2 and Class 3 vessels. This analysis gives an alarm if the DP system is no longer able to maintain position if a worst-case single failure should occur [38, 39]. Failure mode and effects analysis (FMEA) is used during the design and verification phases mainly to review the redundancy of the DP system, and FMEA trials are performed to test this annually [20].

Classification societies in the maritime industry, such as Lloyd's register of shipping, the American Bureau of Shipping, and DNVGL, have traditionally provided support on the quality and safety of vessels and offer notations on DP class with requirements derived from the IMO equipment classes. Class rules are used as criteria for approval and certification and are therefore used by shipyards and vendors to ensure compliance.

A DP vessel that aims to achieve DP equipment class 2 or 3 is usually designed with systems and components distributed into two or several redundancy groups. The redundancy design principle assumes that a worst-case single failure might occur, for example, a failure in a subdivision of the electric power distribution system resulting in loss of the associated generating capacity and the thrusters. Sufficient redundancy then implies that no single failure should result in failures in both subdivisions of the power system (e.g., a common cause failure for all the diesel generators). Hence, each thruster redundancy group need to be able to produce adequate forces in surge, sway, and yaw simultaneously [35].

Guidelines from the International Marine Contractors Association (IMCA) suggest that redundancy should be focused on power generation, management, and distribution, thruster units, position control and reference, and environmental and vessel sensors [40]. Utne et al. [35] show that even though the focus on the technical reliability and redundancy is important, it is insufficient for DP vessels, as systemic hazardous events may be overlooked. With increased complexity of systems and operations, including more software functionality, these approaches may become less adequate in the future.

Today, advanced software-intensive systems onboard a DP offshore vessel are the power system, marine control or automation system, automatic sailing system, DP system, joystick system, manual thruster control system, propulsion and thruster system, equipment packages and systems, safety systems, and auxiliary systems [35, 40]. The DP system is at the focus of this book, as an example of an advanced and complex software intensive system relevant for autonomous marine and maritime systems.

1.4.3 *Incidents in the Offshore Petroleum Industry*

Turret moored Floating Production Storage and Offloading (FPSOs) and Floating Storage Units (FSUs) of the mono-hull type have been used in the North-west European waters since 1986 (Petrojarl I), so far without serious accidents to personnel or the environment in these regions. Although FPSOs are becoming common, operational safety performance may still be considered somewhat unproven, especially when compared to a fixed oil and gas installation. FPSOs and FSUs are serviced by shuttle tankers in the North-west European waters in a tandem offloading configuration. Accidents and incidents involving collision between ST and FPSO/FSU are the main threats, which also have received the highest attention.

Accidents with ST are often initiated by errors induced by combinations of human and organizational factors, technical (design) failures. Effective means to prevent or mitigate the effects of potential operational accidents are thus important.

Five collision accidents occurred in a short 4.5 years period from 1996 to 2000 in UK and Norwegian waters in the North Sea and caused significant concern among the stakeholders. An initial research and development (R&D) project on "FPSO Operational Safety" was conducted in the period 1996–2003 [41–44]. More details are presented by Vinnem et al. [45].

The following are the collision accidents that have occurred worldwide in the period 1996–2019 for the tandem offloading configuration [45]:

- 1996: Emerald FSU, February 28, 1996 [UK)
- 1997: Gryphon FPSO, July 26, 1997 [UK)
- Captain, August 12, 1997 [UK).
- 1998: Schiehallion FPSO, September 25, 1998 [UK)
- 2000: Norne FPSO, March 5, 2000 [N)
- 2006: Njord B FSU, November 13, 2006 [N)
- 2009: Schiehallion FPSO, October 8, 2009 [UK)
- 2012: Offshore Brazil, May 31, 2012 [B).

The latest statistics is presented in a Master of Science (M.Sc.) thesis [46], based on statistics from British and Norwegian sectors. The average United Kingdom (UK) and Norway collision frequency for the period 2004–2013 [46] is the FPSO–shuttle tanker (ST) collision frequency of 1.35 · 10-2 per installation year.

The average collision frequency for the Norwegian sector alone for the period 2004–2013 is virtually the same as the average for the UK and Norwegian sectors [46]. More details are presented by Vinnem et al. [47]. Lundborg [48] analyzed in detail nine accidents and incidents with shuttle tankers on Norwegian Continental Shelf (NCS). It was found that no accidents nor incidents resulted from a single technical failure or human action, while seven out of nine accidents and incidents were due to a combination of technical, human, and organizational failures.

1.5 Shortcomings with Present Works and Future Needs

The main focus for DP vessels has traditionally been on redundancy requirements during the design phase, and the verification of redundancies during commissioning and testing of systems. Too little focus has been on the operations phase [49]. The only focus on the DP operator (DPO) is with respect to training requirements to be licensed as a fully qualified DPO. The scope of attention to the DPO should be much wider [49] and include aspects such as:

- Human factors, such as human–machine interface (HMI), including alarm management and work patterns
- User-centered design of DP system controls.
- Time required for decision-making in critical situations

There is a concern in the Norwegian offshore petroleum industry that DP system incidents and accidents have too high frequency [49] and that improvements are needed. This may suggest that the requirements to DP systems implicitly are higher in the offshore petroleum industry than in the maritime [i.e., "non-offshore"] sector.

Important parts of risk monitoring and risk control of the autonomous system are related to planning of the operation, human–machine interface (HMI), and organizational aspects. To visualize the risk to the human operator and supervisors, indicators may be needed. In general, trending of indicators may contribute to more knowledge and information about a system and/or operation, reducing the uncertainty in risk assessments, and providing useful input into the design and development of such systems.

Indicators can be monitored online or real-time during operation, or they may be sampled on a regular basis, independent of a specific mission. The former is most relevant for online risk assessment. Dong et al. [50] has developed collision risk indicators for autonomous remotely operated subsea vehicles. These indicators can be used to simulate a path and evaluate the associated risk. Hence, the operator may be informed in advance of specific locations along the path where he or she should be on alert in particular or take over operation and maneuvering. An alternative outcome of such a simulation is to select another path.

Fault-tolerant control is an important feature of many engineered systems [51–53]. It enables a system to accommodate the effect of faults. This field is mainly focused on reducing the consequences of internal faults and offers methods for diagnosing whether a behavior is normal or not, whereas the online risk assessment concept is, working to prevent failures related to both human and technical errors. In hindsight it may appear that if signals or warnings had been detected and managed earlier, the undesired event could have been prevented. So, if we use collision risk for an autonomous ship as an example, we may have different collision scenarios, such as powered collision and drifting collision (e.g., loss of position). Different risk models can be developed for the scenarios and the risk modeling must take features of the autonomous ship into consideration, such as energy storage and capacity, path length, speed of vehicle, location, operational aspects, environmental conditions (wind, waves, current), as well as traffic density, path characteristics, visibility, sea temperature, vehicle characteristics, battery management, the technical condition of vehicle, its position, and detection capability. Hence, risk management encompasses a lot more than contingency handling and fault-tolerant control [34].

Summing up, risk models that can provide online decision support including considerations to environmental and operational conditions and limitations, risk scenarios, both proactively and reactively, for safer operation are needed. Proactive means that early warnings on possible violations of the operating envelope are provided, whereas reactively ensures that human operators are given more time for efficient responses and enhanced crisis intervention by providing predictions of possible outcomes. This is the overall topic of this book.

1.6 Objectives and Limitations

The main objective of this book is to present a framework for online risk assessment of automated and autonomous marine and maritime systems. The framework includes multiple modules that model human, hardware, and software behavior, and their interactions over time. In this approach, data flow and possible solution algorithms for each module are presented and discussed.

Multiple case studies are presented to show the applicability and effectiveness of the proposed methodology. The strengths and weaknesses are discussed, and possible solutions to overcome the weaknesses are mentioned.

The framework can be used to provide decision support for human operators or supervisors of autonomous systems, or it can be integrated or utilized by a control system in the autonomous system (supervisory risk control). Hence, the book should be relevant for different audiences, such as industry and system developers, vendors, researchers, and authorities.

DP systems are utilized for a variety of vessels associated with the offshore petroleum industry. Such systems are also used outside this industry, for cruise ships, nuclear power plants, and space shuttles, etc. The applications discussed in the book are limited to the offshore petroleum industry, and mainly the activities where there is a significant major risk associated with the possible failure of DP systems, such as shuttle tankers offloading from FPSO/FSU, mobile drilling units, diving vessels and flotels.

Even with these limitations, it is believed that the framework has a wider potential usage, for instance, in other autonomous systems.

1.7 Overview of Book Structure

In Chap. 2, definitions of important terms related to online probabilistic risk assessment including dynamic, real-time, and online risk assessments are clarified. Methods for online risk assessment such as hazard identification, event sequence diagrams (ESDs), fault tree analysis (FTA), and hybrid methods are presented. In addition, methods for dynamic probabilistic risk assessment of complex systems, including dynamic event sequence diagrams (DESDs), Markov chains, dynamic fault trees, and dynamic Bayesian networks are presented and discussed. Finally, the challenges of available methods are discussed; possible solutions to overcome these challenges are proposed.

Chapter 3 is intended to assist in the data collection and development of online risk assessment for automated marine and maritime systems. The data flow in online risk assessment models is presented. One of the main steps, in the modeling process is data collection and preparation, which is explained in this chapter. Lastly, the quantification process of failure probabilities/frequencies using the collected data is presented and discussed. In this chapter, a dynamic positioning (DP) system as an example of an automated complex system is presented to illustrate the process of data collection and modeling more clearly.

Chapter 4 presents the challenges that complex systems face related to human and organizational factors, hardware, and software interactions and risk quantifications. First, risk assessment of software in automated systems is discussed. A conceptual model to evaluate software risk level is proposed and challenges regarding quantification and cyberattack are presented. Then, risk assessment of human related factors in automated systems are discussed. Finally, a general framework is proposed that considers software and human factors along with hardware elements to perform risk assessment of complex systems. The framework clarifies how the interactions between software, human and hardware could be considered in risk analysis of complex systems.

In Chap. 5, human errors, response time models, and human factors analysis and quantification using different methods are discussed. First, background research on human factor quantifications in risk assessment models is presented. Then, a general overview on the role of operators and their effect on the system failure is discussed. In this chapter, performance shaping factors that affect human behavior are identified. Lastly, human response time models are presented. These models provide a better understanding on the timeline of the decision-making process of operators in emergencies that affect system risk level significantly.

In Chap. 6, a dynamic probabilistic risk model in emergencies is proposed for dynamic positioning systems. The optimization method, discussed in Chap. 2 to generate failure scenarios in a very short execution time, is implemented to dynamic positioning (DP) systems.

In Chap. 7, the online probabilistic risk level of different dynamic positioning (DP) systems is assessed based on the methods presented in Chap. 2. The proposed methods of DPRA are able to provide an online risk level of the system in different operating and environmental conditions that could assist decision makers to make better (safer) decisions.

Lastly, Chap. 8 addresses the main challenges and possible solutions for online risk assessment, related to methodology, data collection, software, uncertainties, complexity and execution time, and application for decision support.

References

1. Schjølberg I, Utne IB (2015) Towards autonomy in ROV operations. IFAC-PapersOnLine 48(2):183–188
2. Hogg T, Ghosh S (2016) Autonomous merchant vessels: examination of factors that impact the effective implementation of unmanned ships. Australian J Marit Ocean Affairs 8(3):206–222
3. Parhizkar T, Mosleh A (2022) Guided probabilistic simulation of complex systems toward rare and extreme events. The 68th Annual Reliability & Maintainability Symposium (RAMS), 2022, Tucson, Arizona
4. Parhizkar T, Roshandel R (2017) Long term performance degradation analysis and optimization of anode supported solid oxide fuel cell stacks. Energy Convers Manage 133:20–30
5. Roshandel R, Parhizkar T (2016) Degradation based optimization framework for long term applications of energy systems. case study: solid oxide fuel cell stacks. Energy 107:172–181
6. Parhizkar T, Hafeznezami S (2018) Degradation based operational optimization model to improve the productivity of energy systems. case study: solid oxide fuel cell stacks. Energy Convers Manage 158:81–91
7. Roshandel R, Parhizgar T (2013) A new approach to optimize the operating conditions of a polymer electrolyte membrane fuel cell based on degradation mechanisms. Energy Syst 4(3):219–237
8. Parhizkar T (2018) Long-term degradation-based modeling and optimization framework. In: Handbook of research on predictive modeling and optimization methods in science and engineering (pp. 192–220). IGI Global
9. Parhizkar T, Mosleh A, Roshandel R (2017) Aging based optimal scheduling framework for power plants using equivalent operating hour approach. Appl energy 205:1345–1363
10. Sotoodeh AF, Parhizkar T, Mehrgoo M, Ghazi M, Amidpour M (2019) Aging based design and operation optimization of organic rankine cycle systems. Energy Convers Manage 199:111892
11. Parhizkar T, Balali S, Mosleh A (2018) An entropy based bayesian network framework for system health monitoring. Entropy 20(6):416
12. Hamed S Nejad, Parhizkar T, Mosleh A (2021) Simulation based probabilistic risk assessment (SIMPRA): risk based design. 31st European Safety and Reliability Conference | 19–23 September 2021, Angers, France
13. Van Hooydonk E (2014) The law of unmanned merchant shipping–an exploration. J Int Marit Law 20(3):403–423
14. Chang YC, Zhang C, Wang N (2020) The international legal status of the unmanned maritime vehicles. Mar Policy, 113:103830
15. Ramos MA, Utne IB, Mosleh A (2019) Collision avoidance on maritime autonomous surface ships: operators' tasks and human failure events. Saf Sci 116:33–44
16. Wahlström M, Hakulinen J, Karvonen H, Lindborg I (2015) Human factors challenges in unmanned ship operations–insights from other domains. Proc Manuf 3:1038–1045
17. Ahvenjärvi S (2016) The human element and autonomous ships. TransNav Int J Mar Navig Safety Sea Transp 10(3)
18. Hegde J, Henriksen EH, Utne IB, Schjølberg I (2019) Development of safety envelopes and subsea traffic rules for autonomous remotely operated vehicles. J Loss Prev Process Ind 60:145–158

19. Thieme CA, Utne IB, Haugen S (2018) Assessing ship risk model applicability to marine autonomous surface ships. Ocean Eng 165:140–154
20. Rokseth B, Utne IB, Vinnem JE (2017) A systems approach to risk analysis of maritime operations. Proc Inst Mech Eng Part O J Risk Reliab 231(1):53–68
21. Banda OAV, Kannos S, Goerlandt F, van Gelder PH, Bergström M, Kujala P (2019) A systemic hazard analysis and management process for the concept design phase of an autonomous vessel. Reliab Eng Syst Safety 191:106584
22. Vinnem JE, Utne IB (2018) Risk from cyberattacks on autonomous ships. Safety Reliab Safe Soc Chang World
23. Tam K, Jones K (2018) Cyber-risk assessment for autonomous ships. In: 2018 international conference on cyber security and protection of digital services (Cyber Security). IEEE, pp 1–8
24. Zhang X, Zhang Q, Yang J, Cong Z, Luo J, Chen H (2019) Safety risk analysis of unmanned ships in inland rivers based on a fuzzy Bayesian network. J Adv Transp
25. Hegde J, Utne IB, Schjølberg I, Thorkildsen B (2018) A Bayesian approach to risk modeling of autonomous subsea intervention operations. Reliab Eng Syst Saf 175:142–159
26. Skogdalen JE, Vinnem JE (2011) Quantitative risk analysis offshore-human and organizational factors. Reliab Eng Syst Saf 96(4):468–479
27. Ramos MA, Thieme CA, Utne IB, Mosleh A (2020) Human-system concurrent task analysis for maritime autonomous surface ship operation and safety. Reliab Eng Syst Safety 195:106697
28. Parhizkar T, Vinnem JE, Utne IB (2019) Dynamic probabilistic safety assessment framework to assist decision-making in complex systems, case study: dynamic positioning drilling unit, Ocean Engineering, 2019, Submitted
29. Utne IB, Schjølberg I, Roe E (2019) High reliability management and control operator risks in autonomous marine systems and operations. Ocean Eng 171:399–416
30. Parhizkar T, Vinnem JE, Utne IB, Mosleh A (2020) Supervised dynamic probabilistic risk assessment of complex systems, Part 1: General Overview. Reliability Engineering & System Safety, 107406
31. Hu Y, Parhizkar T, Mosleh A (2021) Guided Simulation for dynamic probabilistic risk assessment of complex systems: concept, method, and application. Reliability Engineering & System Safety, 108047
32. Parhizkar T, Utne IB, Vinnem JE, Mosleh A (2021) Supervised dynamic probabilistic risk assessment of complex systems, part 2: Application to risk-informed decision making, practice and results. Reliability Engineering & System Safety, 208, 107392
33. National Institute of Standards and Technology (NIST) (2008) Autonomy levels for unmanned systems (ALFUS) Framework, vol I, Terminology, version 2.0, NIST Special Publication 1011-I-2.0
34. Utne IB, Sørensen AJ, Schjølberg I (2017) Risk management of autonomous marine systems and operations. In: International conference on offshore mechanics and arctic engineering, vol 57663, p V03BT02A020. American Society of Mechanical Eng
35. Rokseth B (2018) Safety and verification of advanced maritime vessels-an approach based on systems theory, Doctoral thesis, Marine Department, Norwegian University of Science and Technology
36. Breivik M, Kvaal S, Østby P (2015) From Eureka to K-Pos: dynamic positioning as a highly successful and important marine control technology. IFAC-PapersOnLine 48(16):313–323
37. IMO J (1994) Guidelines for vessels with dynamic positioning systems. MSC/Circular, 645.
38. DNV G (2013) Rules for classification and construction. Germanischer Lloyd SE, Hamburg. Retrieved from: http://rules.dnvgl. com/docs/pdf/gl/maritimerules/gl_i-1-15_e. pdf. Date accessed on 16 June 2021
39. DNV G (2018) RULES FOR CLASSIFICATION, Ships, Part 6 Additional class notations, Chapter 3 Navigation, manoeuvring and position keeping
40. Parhizkar T, Hogenboom S, Vinnem JE, Utne IB (2020) Data driven approach to risk management and decision support for dynamic positioning systems. Reliability Engineering & System Safety, 201, 106964

41. Vinnem JE (2000) Operational safety of FPSOs: initial summary report. OTO Report No.: 2000:086; HSE, 2001
42. Vinnem JE, Hauge S, Huglen Ø, Kieran O, Kirwan B, Rettedal WK et al (2000) Systematic analysis of operational safety of FPSOs reveals areas of improvement. In: SPE international conference on health, safety and environment in oil and gas exploration and production. Society of Petroleum Engineers
43. Vinnem JE, Hokstad P, Saele H, Dammen T, Chen H et al (2002) Operational safety of FPSOs, shuttle tanker collision risk, main report. Trondheim; NTNU; 2002 Oct. Report No.: MK/R 152
44. Vinnem JE, Hokstad P, Dammen T, Saele H, Chen H, Haver S et al (2003) Operational safety analysis of FPSO—shuttle tanker collision risk reveals areas of improvement, OTC paper 15317. OTC conference 2003; Houston, USA
45. Vinnem JE, Røed W (2020) Methodology for quantified risk assessment. In: Offshore risk assessment, vol 2. Springer, London, pp 3–59
46. Lundborg KME (2014) Human technical factors in FPSO-shuttle tanker interactions and their influence on the collision risk during operations in the North Sea (Master's thesis, Institutt for marin teknikk)
47. Vinnem JE, Utne IB, Schjølberg I (2015) On the need for online decision support in FPSO–shuttle tanker collision risk reduction. Ocean Eng 101:109–117
48. Dong Y, Vinnem JE, Utne IB (2017) Improving safety of DP operations: learning from accidents and incidents during offshore loading operations. EURO J Decis Processes 5(1–4):5–40
49. Hogenboom S, Parhizkar T, Vinnem JE (2021) Temporal decision-making factors in risk analyses of dynamic positioning operations. Reliability Engineering & System Safety, 207, 107347
50. Hegde J, Utne IB, Schjølberg I (2016) Development of collision risk indicators for autonomous subsea inspection maintenance and repair. J Loss Prev Process Ind 44:440–452
51. Blanke M, Kinnaert M, Lunze J, Staroswiecki M (2016) Introduction to diagnosis and fault-tolerant control. In: Diagnosis and fault-tolerant control. Springer, Berlin, pp 1–35
52. Parhizkar T, Aramoun F, Saboohi Y (2020) Efficient health monitoring of buildings using failure modes and effects analysis case study: air handling unit system. J Build Eng 29:101113
53. Parhizkar T, Aramoun F, Esbati S, Saboohi Y (2019) Efficient performance monitoring of building central heating system using Bayesian Network method. J Build Eng 26:100835

Chapter 2
Dynamic Probabilistic Risk Assessment

2.1 Probabilistic Risk Assessment

Risk assessment is a systematic process to comprehend the nature of risk, express, and evaluate risk, with the available knowledge [1]. Risk assessment is performed to make better decisions in the design and operation phase [2, 3] of complex systems, resulting in safer conditions [4, 5]. Safety of complex systems, such as nuclear power plants, oil and gas industry, and space shuttles, have been studied for decades [6]. In recent years, the probabilistic and dynamic behavior of complex systems in the risk assessment process has gained increased attention. Sophisticated techniques and methodologies have been developed to analyze different complex systems, and advances have been made in fundamental issues as reviewed in [7, 8].

Risk assessment could be performed qualitatively, quantitatively [9], or (semi-) qualitatively, depending on the aim of study and nature of the event under study. A (semi-) qualitative risk analysis prioritizes the identified risks using a predefined rating scale. Risks can be scored based on their probability or likelihood of occurring and the impact on the study objectives. A quantitative risk analysis is a numerical uncertainty and undesirability assessment. Generally, quantitative risk assessment requires excessive numerical information and may use advanced mathematical and simulation tools to calculate the probabilities and impacts [10].

Probabilistic risk assessment (PRA)[1] is a structured way of quantitative risk assessment to navigate the design and operation of systems toward achieving a certain safety or operational goal. National Aeronautics and Space Administration (NASA) [11] has provided a guideline for PRA for managers and practitioners. In recent years, many studies have performed PRA on different applications, ranging from human health [12] to airports [13] and renewable power plants [14]. Mosleh [15] has argued strengths, limitations, and possible improvements of PRA.

---

[1] Quantitative risk analysis (QRA) in the offshore oil and gas industry and Formal Safety Analysis (FSA) in the maritime industry.

T. Parhizkar et al., *Online Probabilistic Risk Assessment of Complex Marine Systems*, Springer Series in Reliability Engineering,
https://doi.org/10.1007/978-3-030-88098-9_2

The dynamic characteristics and behavior of a system, stochastic processes, operator response times, inspection and testing time intervals, aging of equipment/components, season changes, sequential dependencies of equipment/components, and timing of safety system operation have great influence on system PRA [16, 17]. Conventional PRA methodologies generally have limited capacity in analyzing and quantifying these dynamic characteristics of complex systems. Therefore, it is important to develop methods that can address time-dependent effects in probabilistic risk assessment and provide efficient estimation of risk.

The dynamic PRA (DPRA) has been developed to understand unintended time-dependent interactions of system components, including technical, organizational, and operational factors over time. Different methods have been proposed for considering these interactions of the system in risk assessment [18], of which most are based on the dynamic event sequence diagram/fault tree/Bayesian Network (BN) methodologies [19–21]. In these methods, instead of a unique static event sequence diagram (ESD) or event tree (ET) connected to the specific fault trees (FTs) and BNs, multiple ESDs/ETs/FTs/BNs are developed. At each time step, related ESDs/ETs/FTs/BNs of an incident scenario would be selected based on the environmental and operational characteristics of the system. In this case, the ESDs/FTs/BNs would be updated over time.

2.2 Dynamic, Real-Time, and Online Risk Assessments

Highly automated maritime vessels require system-risk assessment capabilities in operation. According to [22], such capabilities are of vital importance to control the systems safely and efficiently. Decision support tools need to rely on operational risk assessment, providing an analysis of the risk level of a specific operation. In this context, online, real-time, and dynamic terms are used. As an example, [23] proposed an online risk management framework, with the aim of providing online decision support and a real-time risk picture for dynamic positioning system operators. While this paper attempts to explicitly describe terms such as online risk management and dynamic risk assessment, other authors within the field of risk science use the same terms in different ways, and in many cases, without clear definitions. Yang et al. [24] have discussed relationships between operational risk, dynamic risk, real-time risk and have shown that there is significant confusion when it comes to use of consistent terminology.

Interdisciplinary work needs precise terminology to avoid ambiguity. Thus, this section attempts to clarify the definition of online, real-time, and dynamic risk assessments.

2.2.1 *Dynamic Risk Assessment*

The need for more dynamic risk assessment approaches has become more explicit in recent years [25], i.e., risk assessment methods that can be used in operation reflecting any rapid changes or incremental increases in the risk level for standard daily operations [26]. Dynamic means a change in time or any variables through the interaction with the environment. A dynamic risk assessment can be defined as a "method that updates estimated risk of a deteriorating process according to the performance of the control system, safety barriers, inspection and maintenance activities, the human factor, and procedures" [27]. Dynamic risk assessment is the process of assessing risk considering time-varying and/or environmental changes. In other words, a dynamic risk assessment updates the estimated risk of a process according to the changes in the system (e.g., changes in the performance of a control system, safety barriers, inspection and maintenance activities, the human factor, procedure, etc.) [28–35].

Dynamic risk level is defined as the risk level considering the effects of timing and interaction with a changing environment during operation. This definition is in contrast to static risk, where averages and fixed assumptions are used [18, 27, 36].

2.2.2 *Real-Time Risk Assessment*

Oxford Dictionary [37] defines real time as a process "in which input data is processed almost immediately." Expert opinion is that a process is real time, if it is executed immediately while the input is relevant. In condition monitoring systems, real-time data present the most recent and relevant data from the process [38].

Real-time risk level is the risk level being assessed based on the most recent and relevant data. This process is dynamic, and data are updated continuously [39].

Real-time risk assessment is the process of assessing the current and immediate risk level. Available and currently relevant data is used to update an existing risk model in real time [40, 41]. Online risk assessment may be dynamic and include real-time data, as well as other types of data from different sources.

2.2.3 *Online Risk Assessment*

The developments in wireless technology and the availability of cheaper, smaller, and improved sensor technology capable of monitoring assets covering large areas opens up for increased use of online measurements in risk assessment and decision support. Traditionally, risk analysis presents a static picture of the average state of operation or system, based on historical data and expert judgments. For example, the quantitative risk assessment (QRA) in the Norwegian oil and gas industry is typically updated with five-year intervals, whereas safe job analysis (SJA) is carried out regularly in operation.

The purpose of online risk assessment is to support decision-making and give human operators a risk picture and prewarnings of possible deviations in operation. The goal is to improve functionality in safety critical software-based systems, provide risk-based information to operators to reduce the need for manual operation and intervention, and enable longer operator response time if manual intervention becomes necessary [26]. Online risk assessment is based on enhanced utilization of sensors and instrumentations to improve online decision support, as well as the development of new monitoring solutions.

Development of online decision support for safer operation involves risk assessment and modeling, development of data models and representation, inclusion of sensors and communication technology, visual computing and human–machine interface (HMI), human and organizational factors, as well as system integration. The framework, including online risk management and decision support, is illustrated in Fig. 2.1 and consists of a number of modules, such as risk models, data models, HMI, visualization module as well as decision support module [26].

The online risk assessment and risk models build on input data from different sources, i.e., historical data, sensors and measurements, and experience data. The models may be static or include dynamics and may range from empirical models based on historical or online data to physics-based models. Online data is needed to improve the models' effectiveness and the risk monitoring precision level. Interpretation of the data needs mathematical models, both deterministic and stochastic, to enable sufficient information to enable the operators to identify a hazardous situation. The choice depends on factors, such as data stream, computational demand, and uncertainty. Moreover, physics-based models may also be considered as data generators and as basis for simulations and simulator development. The utilization

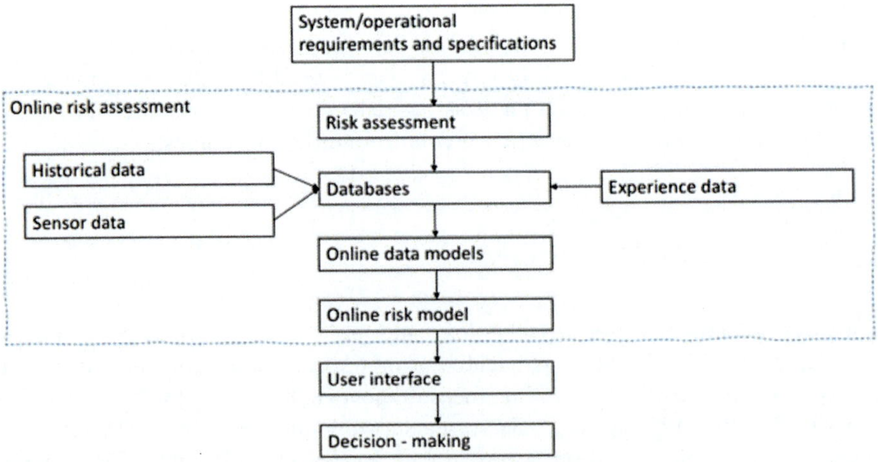

Fig. 2.1 Overall concept for online risk assessment, based on [26]

of data-driven models may enable handling of large data sets, fusing data from different sources, and prediction of information [42]. Experience data from operator training can be used to improve the risk models [26].

2.3 Methods for Online Risk Assessment

The first step in risk modeling is defining the study objective and the system boundary. The establishment of system boundaries for a risk assessment refers to identifying and justifying which affecting parameters are to be included in failure analysis. Figure 2.2 presents a system boundary for the DP system, as an example. The influencing parameters, i.e., environmental and operators' characteristics, and involved technical components (including propulsion, reference, computer, and power systems) have been defined based on the available historical data gathered from the International Marine Contractors Association (IMCA) incident reports. In each incident report, influencing factors and involved components have been identified. Maintenance status is also part of the boundary definition, but this is not available from the IMCA data, and is thus omitted here.

Several incidents/accidents data bases for different industries have been developed. These databases could be used for identification of potential failures and hazardous events in a complex system. Villa et al. [43] presents multiple databases for industries, and has a practical focus, incorporating industry feedback and real-world industry problems and examples.

Using hazard identification, available historical data, and accident investigation reports, incident progression paths (i.e., a sequence of events or path of consequences) can be identified. Incident data may include useful information about initial events and a path of consequences that may result in an incident or accident. Initial events and consequences could be expanded over time using stored online data. Once the

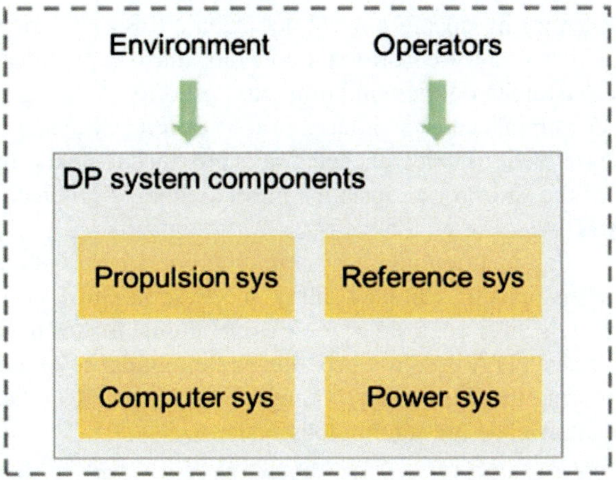

Fig. 2.2 System boundaries of the example dynamic positioning (DP) system [23]

incident progression paths are understood, a feasible method to use may be the hybrid causal logic modeling (HCL) approach. The first step then is to develop event sequence diagrams (ESDs) for scenario quantification, then continue with fault tree analysis (FTA) and Bayesian networks (BN), which are combined to a risk model [44]. The HCL approach is a powerful risk assessment tool for complex systems, where causal relationships are not deterministic. The stochastic causal relationships are quantified using BNs, and results are integrated and propagated into fault trees and event sequence diagrams.

2.3.1 Hazard Identification

A conventional hazard identification (HAZID) method uses brainstorming, generic checklists and taxonomies, as well as accident investigation reports, as a basis for identifying potential hazards and incident progression paths. Autonomous systems, such as DP, consist of complex software, which means that physical separation and redundancy may be overruled by control systems. In addition, the systems may operate and switch between different operational modes with different level of autonomies (LoAs), which introduce hazards related to the human operators' and system controllers' shared interaction and collaboration.

An alternative to HAZID is the systems theoretic process analysis (STPA), which is a hazard identification technique based on the systems-theoretic accident model and processes (STAMP) model [45]. STAMP perceives safety as a control problem, and STPA has been applied in several sectors to identify and analyze hazards, including for dynamic positioning of ships [46, 47] and for autonomous ships [48]. The main outcome of STPA provides a qualitative basis for developing online risk models. Combining STPA with BN risk modeling has been outlined in [48].

In STPA, safety is controlled by enforcing constraints on the system behavior, and accidents happen because of inadequate control. The STPA method consists of defining the purpose of the analysis, modeling the system as a control structure, and identifying unsafe control actions and loss scenarios [49]. Unsafe control actions violate safety constraints, and this must be avoided to prevent accidents. Safety constraints define conditions that need to be in place to prevent hazards and these constraints can be violated in four different ways: a necessary control action is not provided/followed/executed, it is provided too early or too late, it is applied too long or too short, or an unsafe control action is provided [45]. For more information on STPA, see, e.g., [49].

For an autonomous system with operational modes in different LoAs, the mode transitions may impact safety in operation. Supervisors or operators need to be able to take over control or specific operational modes have to be activated when needed. Since STPA does not provide explicit guidance on how to analyze possible failures during these transitions, Yang et al. [50] therefore has proposed an extended STPA tailor-made for autonomous systems.

2.3.2 *Event Sequence Diagrams (ESD)*

An event sequence diagram (ESD) presents event sequences leading to end states. ESDs are able to describe the assumptions used in the related event trees (ET). An event tree extracts the basic event scenario definitions from related ESD and shows this information in a tree structure, which is used to assist in the classification of scenarios based on their consequences and implement a quantification of the scenarios [51]. For more information about ESDs, see, e.g., [52].

An ESD starts from an initial event and then maps out what could happen in the future if components of the complex system fail or succeed in responding appropriately to the initial event. The ESD could be static or dynamic, depending on the system behavior in the specified boundary and the desired level of detail required of the analysis. Details of ESDs in complex systems can be found in [23]. According to the available incident progression paths presented in the IMCA reports, an ESD for the DP system is generated and presented in Fig. 2.3. Initiating events range from bad weather conditions to blackout that results in failure of a component in propulsion, reference, power, and control systems.

2.3.3 *Fault Tree Analysis (FTA)*

In PRA, causes to hazardous events are generally modeled using fault trees. Fault tree is a method for analyzing the failure probabilities of system states using Boolean logic. Fault trees could consider basic events´ interdependencies and the logical behavior of components using logic block diagrams, including multiple events and gates such as AND/OR gates [53]. A detailed description of the concepts and techniques for FTs is presented in [52, 54].

In the example DP system, fault trees of the main subsystems, including propulsion, reference, computer, and power systems have been developed using all available historical incident scenarios from the IMCA reports, as mentioned above. Figure 2.4 presents the propulsion system fault tree of a DP system class 2 [23]. A DP system has

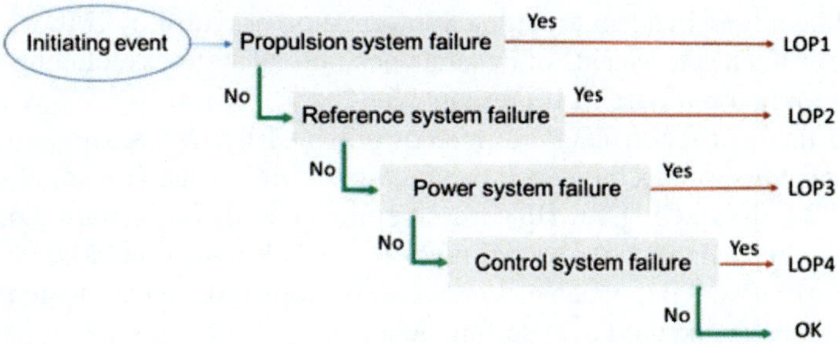

Fig. 2.3 ESD of the example dynamic positioning (DP) system [23], (LOP: loss of position)

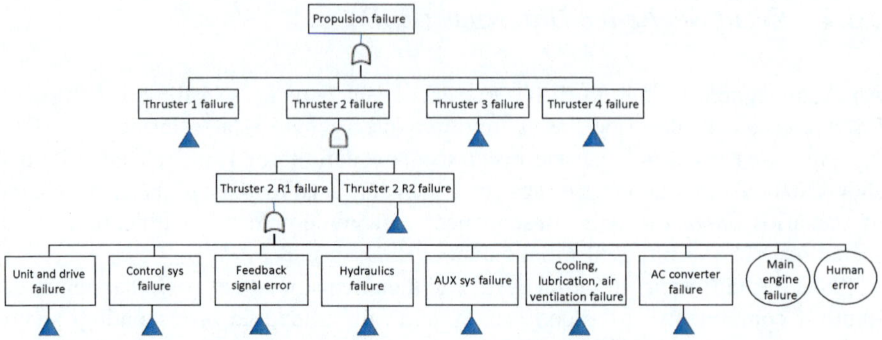

Fig. 2.4 Fault tree of a propulsion system in dynamic positioning (DP) class 2 systems [23], (R1: redundant 1, R2: redundant 2, AUX: auxiliary, AC: alternating current)

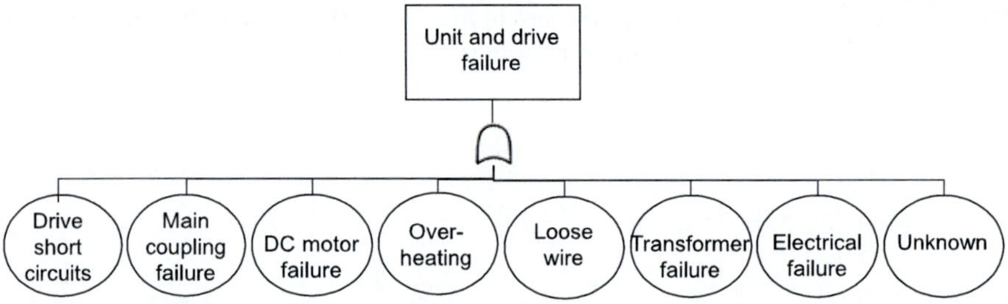

Fig. 2.5 Fault tree of a unit and driver component in a propulsion system, (DC: direct current)

four main thrusters which are presented as Thruster 1–4 in Fig. 2.4. There are redundant units for each thruster in a DP system of Class 2. Each thruster could fail due to technical or human faults and/or failure. The technical failure includes unit and drive fault, control system fault, feedback signal error, hydraulics fault, auxiliary (AUX) system fault, cooling, lubricant and air ventilation system fault, alternating current (AC) converter fault. The fault trees of these components are found in Figs. 2.5, 2.6, 2.7, 2.8, 2.9 and 2.10.

The fault trees presented are examples of fault tree development based on the historical data (i.e., incident report data/information provided by IMCA). The fault tree of other main components of dynamic positioning systems including reference, computer, and power systems is presented in [23].

As the incident report data/information provided by IMCA are gathered from all types of DP systems, the developed fault trees are a generic form of fault trees applicable to all type of DPs. However, the functionality and failure modes of DP systems are dependent on the vessel type. Thus, fault trees should be modified for each type of DP system, which can be found in Appendix A, but the generic forms presented here can be used as a starting point.

The generic fault trees allow for the potential revision and extension of the specific fault trees of the DP system over time. As time passes, different incidents will happen

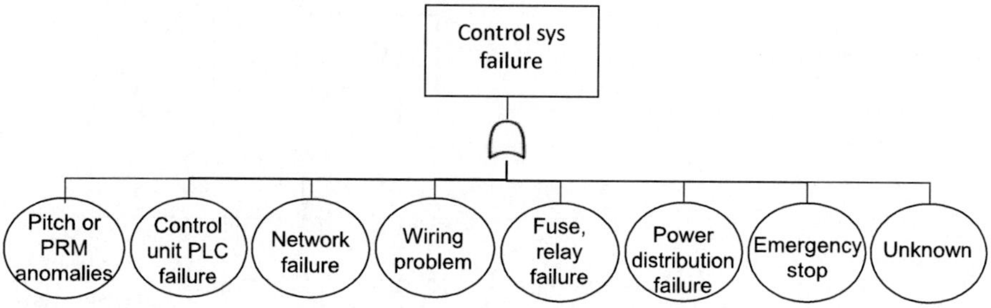

Fig. 2.6 Fault tree of a control component in a propulsion system (PLC: programmable logic controller)

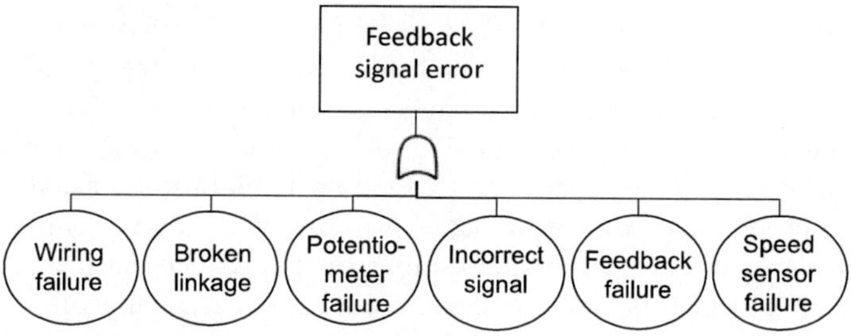

Fig. 2.7 Fault tree of feedback signal function in a propulsion system

Fig. 2.8 Fault tree of cooling, lubrication, and ventilation components in a propulsion system

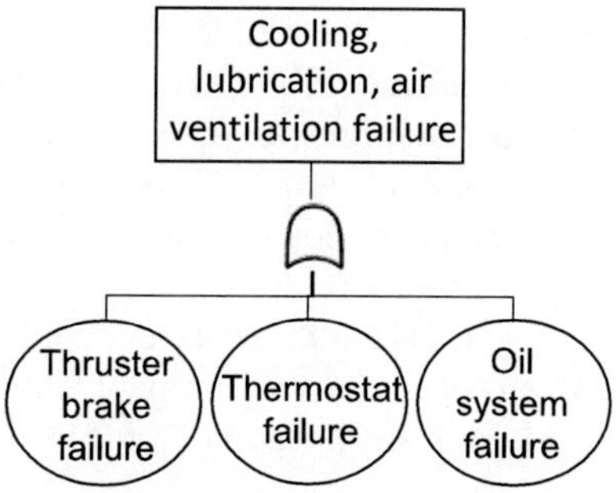

Fig. 2.9 Fault tree of a hydraulic component in a propulsion system

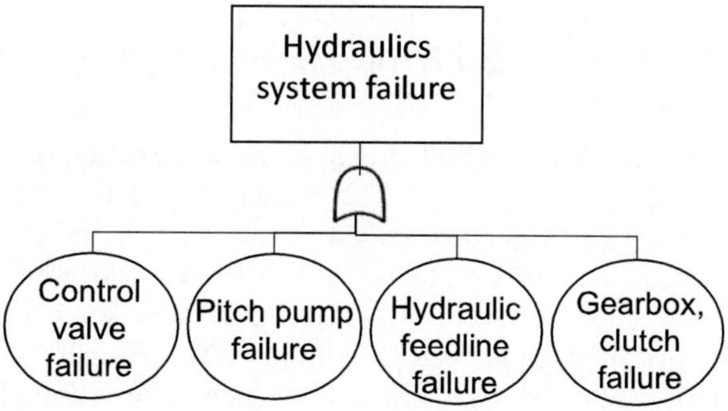

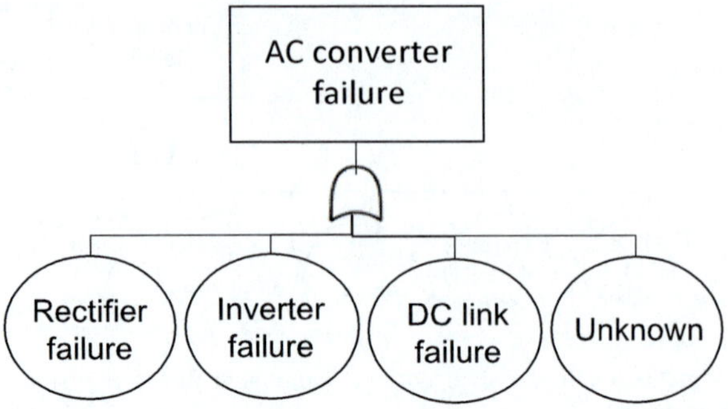

Fig. 2.10 Fault tree of an alternative current (AC) converter fault in a propulsion system (DC: direct current)

in the system that are not included in the specific fault trees. In these cases, the related events in the generic FT could be activated and used to update the specific fault trees of the DP system. For instance, the reference system fault tree in DP diving vessel (Attachment A—Fig. 9.8) has not included software error as one of the basic events. However, according to the generic fault tree the potential for extension in the fault tree exists, i.e., if any software error happens in the DP diving vessel over time, the software error event, with all its connections (as is presented in the generic fault tree) could be activated and added to the reference system fault tree of the DP diving vessel.

In long-term operation, online (sensor) data can be stored and treated as historical data. Then, the stored data can be used to update fault tree/Bayesian network structures at regular intervals. By updating fault trees/BNs and characteristics over time, a risk model would be refined to the system´s particular operating and environmental conditions. In other words, the past behavior of the system then affects the system-risk assessment model. Besides the application in DP systems, this process described here could be applicable to other complex autonomous systems such as cars or space shuttles.

By analyzing the presented fault trees, one of the basic events is human errors. If there is adequate data available, human error could be quantified using BN models. The details of the human error quantification are presented in detail in Chap. 5.

2.3.4 Bayesian Networks

BN analyzes the failure probabilities based on the Bayes theorem applied to a probabilistic graphical model. This graphical model represents a set of variables and their conditional dependencies via directed acyclic graphs (arcs) [55]. Risk-influencing factors and system status are presented as different nodes, and the dependencies between nodes are illustrated using the arcs between nodes.

As an example, Fig. 2.11 shows a BN of operators in a DP system of a drilling unit. At the first step, available incident reports from IMCA (see Chap. 3, Sect. 3.2) of DP systems are studied [24]. The analysis of the incidents from 2004 to 2015

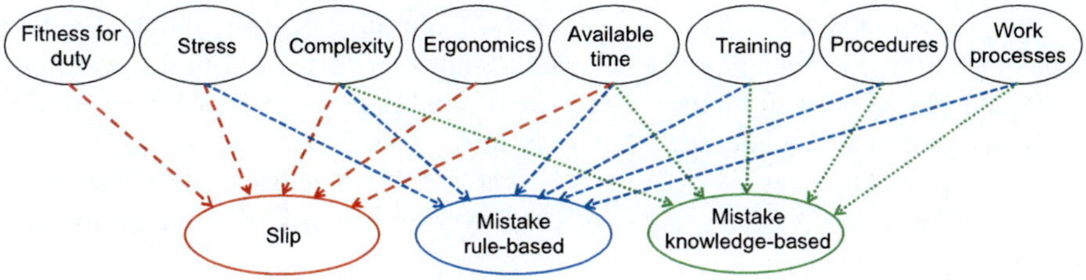

Fig. 2.11 Bayesian network of human error for a DP drilling unit

indicates that human errors could be classified as slip, rule-based and knowledge-based mistakes. These failures are considered as child nodes of a Bayesian network, presented in 2.11. The affecting factors on these types of failures are operators' fitness for duty, stress, complexity, ergonomics, available time, training, procedures, and work processes, which are presented as parent nodes of the Bayesian network (Fig. 2.12). The parent nodes of a BN could take discrete or continuous values. In the presented example, it is assumed that the parent nodes take values of low, medium, and high.

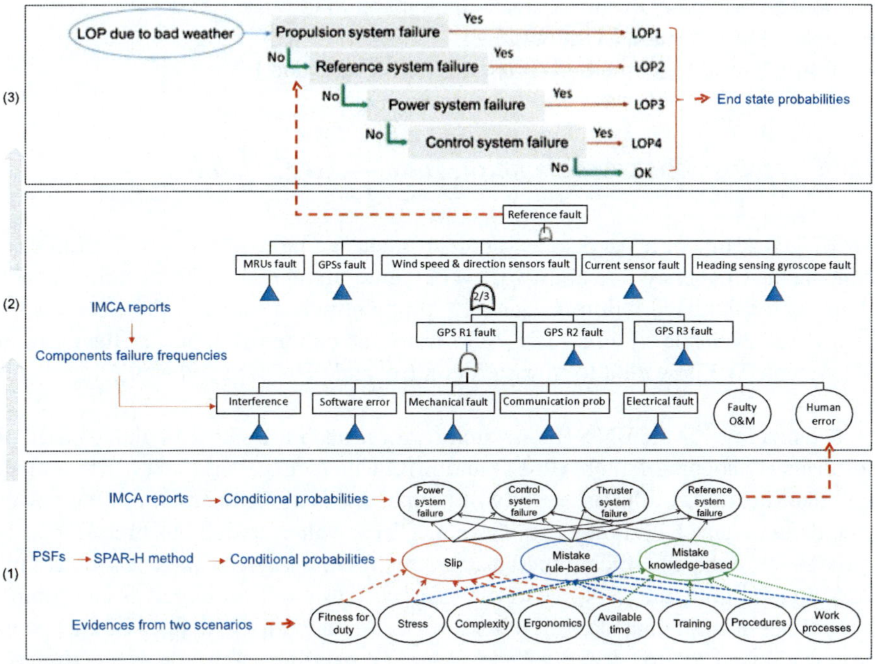

Fig. 2.12 Data flow diagram of a data-driven risk assessment model with the HCL method [23], (LOP: loss of position, MRU: motion reference unit, GPS: global positioning system, R1: redundant 1, R2: redundant 2, R3: redundant 3, O&M: operation & maintenance)

The goal is to calculate the posterior conditional probability distribution of child nodes given the observed evidence of parent nodes. For instance, in this example, we would like to have slip, mistake rule-based, and knowledge-based probabilities for an operator, with a predefined fitness for duty, stress, procedure, work processes, and training level for a task with a specific complexity level and available time. More details on the quantification process of the Bayesian network of human errors are presented in Chap. 5.

2.3.5 Hybrid Methods

A hybrid method refers to a method that combines two or more of the other methods for modeling risk scenarios [56]. The emerging hybrid methods tend to mix fundamentally different representational and computational techniques. The basic concept of PRA is based on the hybrid of event trees and fault trees methods. Hybrid methods help utilizing the advantages of individual methods, resulting in an improved method for risk modeling. These are some of the hybrid methods relevant for dynamic probabilistic risk assessment [15]:

- Linked non-binary event sequence diagram and fault trees
- Linked fault tree and Markov models
- Hybrid causal logic method (linking ET/ESD, FT, and BN).

2.3.6 Hybrid Risk Assessment Model—An Example

An HCL risk model takes data as inputs to Bayesian networks and/or fault trees to evaluate the frequency/probability of event failure. In related ESDs, the initial events along with calculated failure frequencies/probabilities are used to the calculate the online system risk level. Figure 2.12 illustrates the online risk model of the example DP system. The flow data for thruster, control, and power system would be exactly the same.

The first layer is the BNs. Here, human error probabilities are calculated based on performance shaping factors (PSFs) and IMCA incident reports´ data. The outputs are quantifications of the probability of human errors in the operation of the power, propulsion, control, and reference systems. These values and failure frequencies of systems' components update the basic events probabilities/frequencies in all fault trees. In the figure, the fault tree of the reference system is presented as an example (the full model still considers all other fault trees, including fault trees of the power, propulsion, and control system). Then, failure probabilities of the top events from the fault trees are taken as an input by the event sequence diagram (ESD), and according to the initial event, the end state probabilities are calculated.

2.4 Dynamic Probabilistic Risk Assessment Methods

While shortcomings of conventional PRA techniques have been recognized, questions have been raised regarding the capacity of the analysis methods to completely represent the interactions between the hardware, software, and human behavior [2]. Moreover, the order of the events does not rely solely on the causal relationship between them, but also on their timing and the magnitude of the process variables at the time of the failure.

Dynamic probabilistic risk assessment (DPRA) addresses the dynamic behavior of complex systems in the process of risk assessment by a systematic modeling approach. As opposed to conventional PRA methods that do not consider dynamic behavior and study the system in a specific timeframe in an "isolated" environment, DPRA tries to understand complex and dynamic relations between system components. DPRA methods or simulation-based PRA aim at estimating the system risk level, including operational and environmental uncertainties. A simulation engine is used that generates branches at specific time steps with associated probabilities for the risk scenarios, e.g., the accident dynamic simulator (ADS), analysis of dynamic accident progression trees (ADAPT), and risk analysis virtual environment (RAVEN) [57]. Advantages are that they can more realistically represent event sequence timing, provide more insights into condition assessment and consequences, aid in scenario generation, and include operator responses. Simulation-based methodologies are expected to play a critical role in risk assessment of autonomous systems to assist in the detailed understanding of the systems and operations, to collect data, and to simulate outcomes of different potential risk scenarios. The next sections explore the primary techniques used to conduct DPRA of complex systems, developed in recent years. An enhanced framework of ADS is further presented in Chap. 4.

2.4.1 Dynamic Event Sequence Diagram (ESD)/Dynamic Event Trees (ET)

Dynamic ESDs and ETs are methods for evaluating occurrence probability of end states, considering the dynamic behavior of the system. In these methods, time of event occurrence, the order of event occurrence, dependency rules, hierarchy identification, and constraints/boundaries events interdependencies are defined by multiple type of events, conditions, and logical gates. Tables 2.1 and 2.2 present the main events and conditions/gates in dynamic ESD and ET analysis, respectively [58].

Table 2.1 Event definitions in dynamic event sequence diagram/event tree [58]

Term	Definition
Deterministic delays	It represents a situation where no event can occur in a scenario for the duration of the delay
Random delays	This is an event/delay that would occur with certainty within a time window. The time of occurrence of the event or the extent of the delay is, however, a random variable

Table 2.2 Types of conditions and gates in dynamic event sequence diagram/event tree [58]

Term	Symbol	Definition
Time conditions	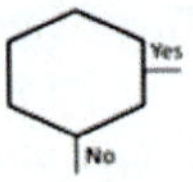	They are based on deterministic time limits and represent a condition of the form a < t < b. They lead to two outcomes depending on whether the condition is met or not
Competition conditions	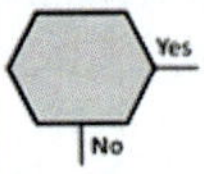	They are used to model a "race" between two processes/events. They lead to two outcomes depending on which process/event is completed first
Physical variable conditions	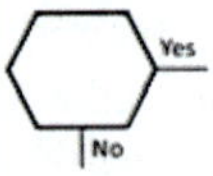	They are based on deterministic process variable limits and present a condition of the form a < p < b. They lead to two outcomes depending on whether the condition is met or not
Output OR gate	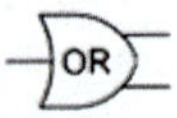	This is used to model multiple mutually exclusive outcomes. This gate has one input and multiple outputs
Input OR gate	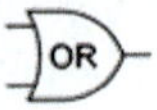	Events from different paths leading to the same event are combined using as input OR gate. This gate has multiple inputs but a single output
Output AND gate	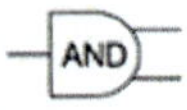	Multiple concurrent independent processes can be modeled using this gate. This gate has one input and multiple outputs
Inputs AND gate		Situation where scenarios cannot progress until a certain set of independent scenarios have occurred (synchronizations) and can be described using the input AND gate. This gate has multiple inputs and a single output

2.4.2 Markov Chain (Continuous Event Tree)

Markov chain, also known as continuous event tree, is a chain of states including the probabilities for transitioning between these states. This method is mostly used to characterize systems that cannot be modeled using closed-form solutions and reliability block diagrams. In DPRA, this method could be utilized to present the dependency and the probabilities for transitioning of states over time [59] (Fig. 2.13).

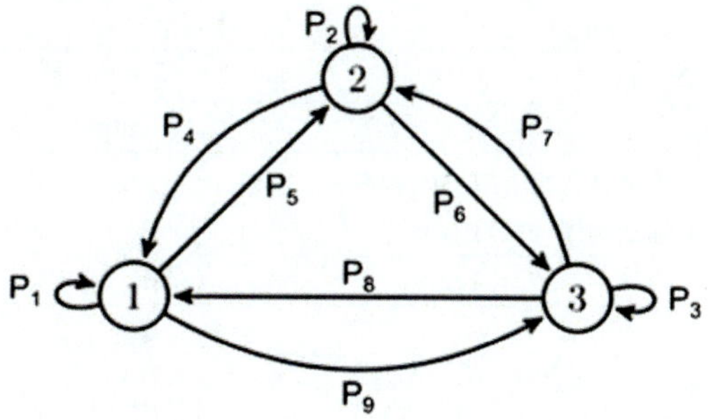

$$\begin{matrix} Transition \\ matrix \end{matrix} = \begin{bmatrix} P_1 & P_5 & P_9 \\ P_4 & P_2 & P_6 \\ P_8 & P_7 & P_3 \end{bmatrix}$$

Fig. 2.13 A sample Markov chain model as an example and its related transition matrix

2.4.3 Dynamic Fault Tree

In dynamic fault trees, the defined events and gates in static fault trees are expanded to consider the dynamic behavior of the system over time. In dynamic simulation, as time passes, interconnected contributory events and conditions are updated. Time-dependent gates are defined to consider time dependency (dynamic behavior) in the fault tree model [60]. Table 2.3 presents some of the time-dependent gates used in dynamic fault tree models.

In Table 2.3, the numbers of 1, 0, and imp represent failure, not failure, and impossible status of the events A, B, and C, respectively.

PAND gate reaches a fault state if all of its input components have failed in a preassigned order. As shown in Table 2.3, when the first component failed followed by the second component, it is identified as failure. However, when the second component has failed first followed by the first component, it is not considered as failure.

SEQ gate is similar to PAND gate; however, occurrence of events is forced to take place in a specific way. No components can fail prior to previous components.

SPARE gate has one active component (event A) and several spare components (event B). SPARE gate reaches a fault state if all active and spare components fail simultaneously.

FDEP gate has a trigger event (event C) and several dependent events (event A, B). If trigger event fails, dependent events will fail accordingly. However, failure of dependent events would not result in failure of trigger event.

2.4.4 Dynamic Bayesian Network (DBN)

In dynamic simulation, variables and their interdependencies in a BN are updated over time. As a result, the dynamic BN is introduced that relates variables to each other over adjacent time steps [62]. Figure 2.14 presents a dynamic BN; as can be seen, the probability of nodes depends on some nodes in time T-1; in addition to conditional probabilities in time T.

Table 2.3 Types of logic gates in dynamic fault trees [61]

Definition	Symbol	Truth table			
		A	B	C	Output
PAND	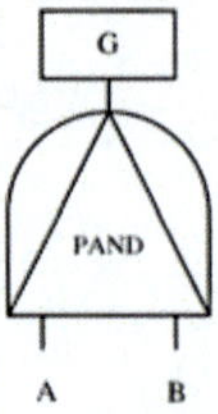	1[a] (first)	1 (second)	–	1
		1 (second)	1 (first)	–	0
		0	1	–	0
		1	0	–	0
		0	0	–	0
SEQ	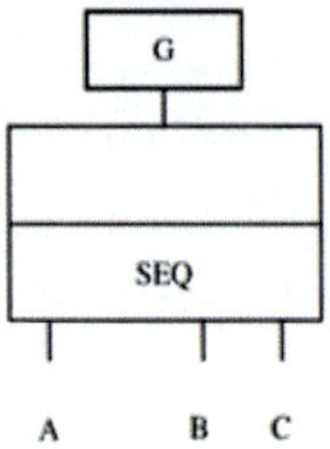	0	0	0	0
		0	0	1	Imp[b]
		0	1	0	Imp
		0	1	1	Imp
		1	0	0	0
		1	0	1	Imp
		1	1	0	0
		1	1	1	1
SPARE	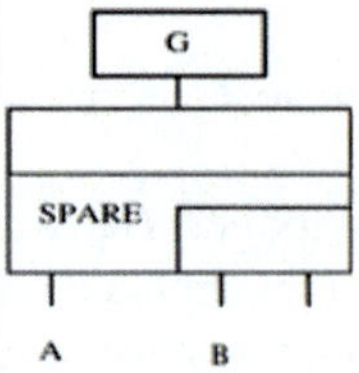	1	1	–	1
		0	1	–	0
		1	0	–	0
		0	0	–	0
FDEP	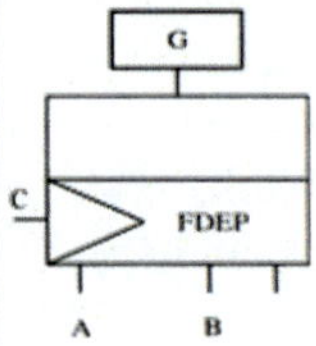	1	1	1	1
		0/1	0/1	0	0

[a] Failure status
[b] Impossible

2.5 Advanced Online Dynamic Probabilistic Risk Assessment Methods

The future of risk assessment is presented in [63]. A main challenge is related to the development of the DPRA for a complex system. Risk assessment is well established for cases with considerable data and system behavior information. In this regard, multiple statistical and probabilistic tools can be used to provide valuable information for decision support in many types of applications.

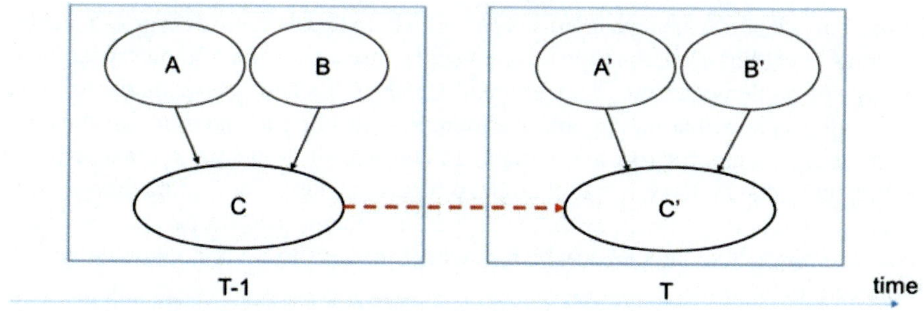

Fig. 2.14 A sample dynamic Bayesian Network (DBN) as an example

However, a risk-informed decision-making process is primarily regarding situations characterized by large uncertainties. These situations necessitate the assessment of multiple alternative possible scenarios and, occasionally, unknown scenarios. It is a key challenge for the risk field to develop frameworks for this purpose. Chang et al. [16] performed an in-depth study on this concept, in particular the effect of human factor uncertainty over time in risk assessment. According to their study, possible alternative scenarios increase with uncertainties and time.

The accuracy and execution time of a DPRA model is highly dependent on the number of predicted possible scenarios. A limited number of predicted scenarios results in an inaccurate model of a system. Increasing the number of predicted scenarios, enhance the accuracy of the model. However, in some cases, it is so complicated and sometimes impossible to consider all probable scenarios. In the case of predicting all possible scenarios, the number of scenarios could growth dramatically over time requiring high execution time. High execution time is against the nature of dynamic probabilistic risk assessment. In most of the studies, the results should be available in real time and operators should be able to use the results to make better decisions, both in normal operation, but also in emergency situations.

Currently, there are two foreseeable main approaches to confront this challenge: One approach is to utilize predictive models in the process of risk assessment and try to predict changes in system behavior and operation impacting risk, for example, based on machine learning. The other approach is to use optimization methods to generate more desirable scenarios, instead of generating all possible scenarios that needs high execution time. These two approaches, which also might be combined, are presented in Sects. 2.5.1 and 2.5.2.

2.5.1 Predictive Models Based on Machine Learning Methods

Artificial intelligence and machine learning for risk assessment is an emerging area through data-driven inputs [64]. In [64], a review on machine learning methods, utilized in risk assessment problems is performed. In this review, the type of machine learning algorithm used, the type of input source used, the type of industry targeted, the type of implementation, and the intended risk assessment phase are determined. The findings show that the automotive industry is leading the adoption of machine learning algorithms for risk assessment. In addition, it is mentioned that artificial neural networks are the most applied data-driven machine learning method used in risk assessment. A challenge with these types of methods in a risk perspective is that they are "black box" methods, which makes it difficult to track how outcomes are produced in the model.

Generally, the increased availability of sensor data and improved computational capability provide enhanced opportunities for dynamic risk assessments [63]. Risk-informed decision-making has to deal with, to an increasing extent, unknown situations, such as incidents in a self-learning complex system. In self-learning systems, the behavior of the system and its responses to the environment are updated/changed over time, e.g., the system response over time to a specific incident will not be the same as its response in the beginning of its operation.

In these cases, system behavior and principal rules are updated continuously. Therefore, the consequences and recovery plans of an incident will change over time. The available probabilistic dynamic risk approaches and methods are not capturing self-learning behavior of complex systems. Although risk assessment methods evaluate the risk level of the system, the predictive methods need to be extended to assist in predicting self-learning system behavior over time. In this regard, predictive methods and analysis in self-learning systems can be considered new approaches for improving DPRA methods. For more on machine learning, see Sect. 3.3.

2.5.2 Optimization Methods

In this section, an optimization method to assist DPRA models is introduced. The proposed method is applicable to DPRA modeling methods, such as the dynamic fault tree, dynamic Bayesian network, dynamic event sequence diagram, and hybrid methods.

This method can generate and sort failure scenarios in a short execution time. Hence, this method may be highly valuable for risk-informed decision-making in emergencies and situations with time restrictions. The method is able to generate failure scenarios after an incident, considering operation and environmental conditions. In this approach, all alternative possible scenarios need not be solved; instead, failure scenarios are derived using an optimization model. To sort the failure scenarios without considering all possible scenarios, an optimization concept is introduced. An optimization model is developed based on the system's characteristic and constraints to obtain more probable scenarios. In this section, the optimization model and related solution algorithms are presented.

An optimization problem typically has three main elements: The first is an objective function that is to be maximized or minimized according to the study objective [65–70].

The objective of this study is to maximize the probability of the scenario occurrence.

$$\text{Maximize occurrence probability} \tag{2.1}$$

At the first step, the model obtains the most probable scenario; subsequently, this scenario is eliminated, and the model is executed again to obtain the second most likely scenario. The model is executed until all sorted failure scenarios are derived.

The objective function can obtain scenarios with higher risk levels as well. In this case, the objective function is equal to the risk level evaluation. As presented in Eq. (2.2), in addition to scenario probabilities, scenario consequences should be considered in the objective function.

$$\text{Maximize (occurrence probability} \times \text{consequence)} \tag{2.2}$$

In this case, the model obtains a scenario with the highest risk level; subsequently, this scenario is eliminated, and the model is executed again to obtain the next scenario with the highest risk level. The model is executed until all failure scenarios are sorted based on the risk level.

The second element in an optimization model is decision variables that should be optimized to maximize and/or minimize the objective function. Two types of continuous and discrete decision variables exist [65]. The decision variables in this study are the number and type of events in a failure scenario. These variables are discrete and can be defined as integer variables in the optimization model.

$$x \in \{0, 1\} \tag{2.3}$$

where 1 and 0 represent the existence and non-existence of a connection or event in an event sequence diagram, respectively.

The third element of an optimization problem is a set of constraints, which are restrictions on the values that the variables can assume [65]. The constraints should be defined according to the system under study. A system dynamic model and all applicable logical rules should be considered. The constraints can be linear or nonlinear functions. Most risk assessment studies have nonlinear constraints. Equations (2.4–2.6) present some constraints that can be considered in risk assessment problems.

$$\begin{array}{l}\text{Logical constraints:}\\ -\text{Human cognition rules.}\end{array} \tag{2.4}$$

Principle-based constraints:
– Derived from complex systems imulation model (2.5)
– Principal rules of probability and risk assessment.

System study specification and limitations:
– Riskcriteria (2.6)
– Standards that should be followed.

The main challenge of using the optimization method is the selection of the optimization algorithm. Within a broad optimization model, the mathematical properties of risk assessment problems include modeling nonlinear equations considering integer decision variables.

The mathematical properties imply that risk assessment optimization problems can be classified as mixed-integer nonlinear programming. There are multiple solution algorithms applicable to mixed-integer nonlinear programming problems. According to the nonlinearity level of the system and type of decision variables (binary, integer, continuous) the best solution algorithm should be defined.

This method is applied to a case study in Chap. 6 (Sect. 6.2). In the case study, the objective is to find more probable failure scenarios in a DP system. An emergency situation is presented as an example. An optimization algorithm based on the model nonlinearity and type of the decision variable is selected and applied to the case study. Using the optimization model, more probable failure scenarios for the case under study are generated. The results show that the optimization method reduces the computational execution time significantly which is valuable for decision support in operation. Other methods and approaches may also be feasible in this respect and are currently under investigation; see [71].

References

1. Gerba CP (2019) Risk assessment. In: Environmental and pollution science. Academic Press, pp 541–563
2. Parhizkar T, Mosleh A, Roshandel R (2017) Aging based optimal scheduling framework for power plants using equivalent operating hour approach. Applied energy, 205, 1345–1363
3. Sotoodeh AF, Parhizkar T, Mehrgoo M, Ghazi M, Amidpour M (2019) Aging based design and operation optimization of organic rankine cycle systems. Energy Convers Manage 199:111892
4. Parhizkar T, Balali S, Mosleh A (2018) An entropy based bayesian network framework for system health monitoring. Entropy 20(6):416
5. Hamed S Nejad, Parhizkar T, Mosleh A (2021) Simulation Based Probabilistic Risk Assessment (SIMPRA): risk Based Design. 31st European Safety and Reliability Conference | 19-23 September 2021, Angers, France
6. Parhizkar T, Utne IB, Vinnem JE, Mosleh A (2021) Dynamic probabilistic risk assessment of decision-making in emergencies for complex systems, case study: dynamic positioning drilling unit. Ocean Eng 237:109653

7. Parhizkar T, Vinnem JE, Utne IB, Mosleh A (2020) Supervised dynamic probabilistic risk assessment of complex systems, Part 1: General Overview. Reliability Engineering & System Safety, 107406
8. Parhizkar T, Utne I.B, Vinnem JE, Mosleh A (2021) Supervised dynamic probabilistic risk assessment of complex systems, part 2: Application to risk-informed decision making, practice and results. Reliability Engineering & System Safety, 208:107392
9. Hogenboom S, Parhizkar T, Vinnem JE (2021) Temporal decision-making factors in risk analyses of dynamic positioning operations. Reliability Engineering & System Safety, 207:107347
10. Parhizkar T, Mosleh A (2022) Guided probabilistic simulation of complex systems toward rare and extreme events. The 68th Annual Reliability & Maintainability Symposium (RAMS), Tucson, Arizona
11. Stamatelatos M, Dezfuli H, Apostolakis G, Everline C, Guarro S, Mathias D, Youngblood R et al (2011) Probabilistic risk assessment procedures guide for NASA managers and practitioners
12. Barrio-Parra F, Izquierdo-Díaz M, Dominguez-Castillo A, Medina R, De Miguel E (2019) Human-health probabilistic risk assessment: the role of exposure factors in an urban garden scenario. Landsc Urban Plan 185:191–199
13. Parhizkar T, Aramoun F, Esbati S, Saboohi Y (2019) Efficient performance monitoring of building central heating system using Bayesian Network method. J Build Eng 26:100835
14. Parhizkar T, Aramoun F, Saboohi Y (2020) Efficient health monitoring of buildings using failure modes and effects analysis case study: air handling unit system. J Build Eng 29:101113
15. Hu Y, Parhizkar T, Mosleh A (2021) Guided simulation for dynamic probabilistic risk assessment of complex systems: concept, method, and application. Reliab Eng Syst Saf. p 108047
16. Chang YHJ, Mosleh A (2007) Cognitive modeling and dynamic probabilistic simulation of operating crew response to complex system accidents: part 1: overview of the IDAC model. Reliab Eng Syst Saf 92(8):997–1013
17. Chang YHJ, Mosleh A (2007) Cognitive modeling and dynamic probabilistic simulation of operating crew response to complex system accidents. Part 2: IDAC performance influencing factors model. Reliab Eng Syst Saf 92(8):1014
18. Villa V, Paltrinieri N, Khan F, Cozzani V (2016) Towards dynamic risk analysis: a review of the risk assessment approach and its limitations in the chemical process industry. Saf Sci 89:77–93
19. Nasif H, Kawai K (2019) Dynamic event tree tool for the safety assessment of a nuclear power plant. In: The proceedings of the international conference on nuclear engineering (ICONE) 2019.27. The Japan Society of Mechanical Engineers, p 2032
20. Wang J (2019) A dynamic fault tree approach for time-dependent logical modeling of autonomous flight systems (No. 2019-01-1358). SAE Technical Paper
21. Levy R (2014) Dynamic Bayesian network modeling of game based diagnostic assessments. CRESST report 837. National center for research on evaluation, standards, and student testing (CRESST)
22. Thieme CA, Utne IB (2017) Safety performance monitoring of autonomous marine systems. Reliab Eng Syst Saf 159:264–275
23. Parhizkar T, Hogenboom S, Vinnem JE, Utne IB (2020) Data driven approach to risk management and decision support for dynamic positioning systems. Reliab Eng Syst Saf 201:106964
24. Yang X, Haugen S, Paltrinieri N (2018) Clarifying the concept of operational risk assessment in the oil and gas industry. Saf Sci 108:259–268
25. Skogdalen JE, Vinnem JE (2011) Quantitative risk analysis offshore—human and organizational factors. Reliab Eng Syst Saf 96(4):468–479
26. Vinnem JE, Utne IB, Schjølberg I (2015) On the need for online decision support in FPSO–shuttle tanker collision risk reduction. Ocean Eng 101:109–117
27. Khan F, Hashemi SJ, Paltrinieri N, Amyotte P, Cozzani V, Reniers G (2016) Dynamic risk management: a contemporary approach to process safety management. Curr Opin Chem Eng 14:9–17

28. Yang X, Mannan MS (2010) The development and application of dynamic operational risk assessment in oil/gas and chemical process industry. Reliab Eng Syst Saf 95(7):806–815
29. Bhatia K, Khan F, Patel H, Abbassi R (2019) Dynamic risk-based inspection methodology. J Loss Prev Process Ind 62:103974
30. Hsueh KS, Mosleh A (1996) The development and application of the accident dynamic simulator for dynamic probabilistic risk assessment of nuclear power plants. Reliab Eng Syst Saf 52(3):297–314
31. Abimbola M, Khan F, Khakzad N (2014) Dynamic safety risk analysis of offshore drilling. J Loss Prev Process Ind 30:74–85
32. Bouloiz H, Garbolino E, Tkiouat M, Guarnieri F (2013) A system dynamics model for behavioral analysis of safety conditions in a chemical storage unit. Saf Sci 58:32–40
33. Kalantarnia M, Khan F, Hawboldt K (2009) Dynamic risk assessment using failure assessment and Bayesian theory. J Loss Prev Process Ind 22(5):600–606
34. Landucci G, Paltrinieri N (2016) Accident frequency evaluation to support dynamic risk studies. Chem Eng Trans 48
35. Paltrinieri N, Khan F (eds) (2016) Dynamic risk analysis in the chemical and petroleum industry: evolution and interaction with parallel disciplines in the perspective of industrial application. Butterworth-Heinemann, Oxford, UK
36. Villa V, Paltrinieri N, Cozzani V (2015) Overview on dynamic approaches to risk management in process facilities. Chem Eng Trans 43:2497–2502
37. Oxford Dictionary (2012) Oxford english dictionary, 7th edn. Oxford University Press, Oxford, UK
38. Kelly SDT, Suryadevara NK, Mukhopadhyay SC (2013) Towards the implementation of IoT for environmental condition monitoring in homes. IEEE Sens J 13(10):3846–3853
39. Harder C, Sutler J, Sawaryn S, Hale NW (March 2015) Real time data monitoring experience results in enhanced safety and efficiency. In: SPE/IADC drilling conference and exhibition. Society of petroleum engineers
40. Zhou J (2010) SPA–fuzzy method based real-time risk assessment for major hazard installations storing flammable gas. Saf Sci 48(6):819–822
41. Hosseini SM, Abdollahi R, Karrari M (2018) Inclusive design and implementation of online load angle measurement for real-time transient stability improvement of a synchronous generator in a smart grid. IEEE Trans Industr Electron 65(11):8966–8972
42. Foss B (2012) Process control in conventional oil and gas fields—challenges and opportunities. Control Eng Pract 20(10):1058–1064
43. Rausand M, Barros A, Høyland A (2020) System reliability theory: models, statistical methods, and applications. John Wiley & Sons
44. Wang C (2007) Hybrid causal logic methodology for risk assessment. Doctoral dissertation, Department of Mechanical Engineering, University of Maryland.
45. Leveson NG (2016) Engineering a safer world: systems thinking applied to safety. The MIT Press, p 560
46. Rokseth B, Utne IB, Vinnem JE (2018) Deriving verification objectives and scenarios for maritime systems using the systems-theoretic process analysis. Reliab Eng Syst Saf 169:18–31
47. Rokseth B, Utne IB, Vinnem JE (2017) A systems approach to risk analysis of maritime operations. Proc Inst Mech Eng, Part O: J Risk Reliab 231(1):53–68
48. Utne IB, Rokseth B, Sørensen AJ, Vinnem JE (2020) Towards supervisory risk control of autonomous ships. Reliab Eng Syst Saf 196:106757
49. Leveson NG, Thomas JP (2018) STPA handbook. MA, USA, Cambridge
50. Yang X, Sandøy S, Utne IB, Rokseth B, Ramos M (2020) A systems-theoretic approach to hazard identification of marine systems with dynamic autonomy. Ocean Eng 217:107930
51. Zhou J, Reniers G, Khakzad N (2016) Application of event sequence diagram to evaluate emergency response actions during fire-induced domino effects. Reliab Eng Syst Saf 150:202–209
52. Rausand M (2011) Risk assessment: theory, methods, and applications, vol 115. Wiley, New York, USA.

53. Vesely WE, Goldberg FF, Roberts NH, Haasl DF (1981) Fault tree handbook. Nuclear Regulatory Commission, Washington DC
54. Fussell JB (1976) Fault tree analysis: concepts and techniques. In: Pressure vessels and piping: design and analysis, IV
55. Yang L, Lee J (2012) Bayesian Belief Network-based approach for diagnostics and prognostics of semiconductor manufacturing systems. Robot Comput-Integr Manuf 28(1):66–74
56. Madden BP (1990) The hybrid model for concept development: its value for the study of therapeutic alliance. Adv Nurs Sci
57. Diaconeasa MA (2017) Integration of qualitative and quantitative hybrid causal logic into a simulation-based platform for probabilistic risk assessment of nuclear power plants. Ph.D. thesis, University of California, Los Angeles
58. Swaminathan S, Smidts C (1999) The event sequence diagram framework for dynamic probabilistic risk assessment. Reliab Eng Syst Saf 63(1):73–90
59. Brooks S, Gelman A, Jones G, Meng XL (eds) (2011) Handbook of Markov chain Monte Carlo. CRC Press, Florida, USA
60. Gascard E, Simeu-Abazi Z (2018) Quantitative analysis of dynamic fault trees by means of Monte Carlo simulations: event-driven simulation approach. Reliab Eng Syst Saf 180:487–504
61. Dugan JB, Bavuso SJ, Boyd MA (1992) Dynamic fault-tree models for fault-tolerant computer systems. IEEE Trans Reliab 41(3):363–377
62. Sterritt R, Marshall AH, Shapcott CM, McClean SI (October 2000) Exploring dynamic Bayesian belief networks for intelligent fault management systems. IEEE international conference on systems, man and cybernetics. 'Cybernetics evolving to systems, humans, organizations, and their complex interactions', vol 5, pp 3646–3652
63. Zio E (2018) The future of risk assessment. Reliab Eng Syst Saf 177:176–190
64. Hegde J, Rokseth B (2020) Applications of machine learning methods for engineering risk assessment–a review. Saf Sci 122:104492
65. Loucks DP, van Beek E (2017) An introduction to optimization models and methods. In: Water resource systems planning and management. Springer, New York, USA, pp. 93–177
66. Parhizkar T, Roshandel R (2017) Long term performance degradation analysis and optimization of anode supported solid oxide fuel cell stacks. Energy Convers Manage 133:20–30
67. Roshandel R, Parhizkar T (2016) Degradation based optimization framework for long term applications of energy systems, case study: solid oxide fuel cell stacks. Energy 107:172–181
68. Parhizkar T, Hafeznezami S (2018) Degradation based operational optimization model to improve the productivity of energy systems, case study: solid oxide fuel cell stacks. Energy Convers Manage 158:81–91
69. Roshandel R, Parhizgar T (2013) A new approach to optimize the operating conditions of a polymer electrolyte membrane fuel cell based on degradation mechanisms. Energy Syst 4(3):219–237
70. Parhizkar T (2018) Long-term degradation-based modeling and optimization framework. In: Handbook of research on predictive modeling and optimization methods in science and engineering (pp. 192–220). IGI Global
71. Maidana RG, Parhizkar T, Thieme CA, Ramos M, Utne I, Mosleh A (2021) Towards risk-based autonomous decision-making with accident dynamic simulation, ESREL conference, France

Chapter 3
Data Sources and Development for Online Risk Quantification

3.1 Data Flow of Online Risk Assessment

Online risk assessment models utilize different sources of data. Historical or empirical data could be gathered from system manufacturers, operators, and experts, etc. Real-time data is all data that is delivered immediately after collection, e.g., from system sensors and actuators. As time passes, the collected data could be stored and used as system historical data. There may be multiple data-driven models involved in online risk assessment [1].

Figure 3.1 presents the possible flow of data for online risk assessment. At the data collection layer, historical and real-time data about the system should be identified and gathered. These data are stored in a database and as presented in the figure, may include:

- Environmental, operational, and technical data about the system under study
- Human and organizational factors that affect system operation
- Incident details such as initial events, root causes, and consequences
- Maintenance types and schedules of the systems' components.

In the database, the amount of stored data and data that should be executed in real time are managed. For complex systems, often the number of incidents or historical data are limited, and simulation methods can be utilized to generate data for a database [2], i.e., if the number of available incidents or information about potential failures are limited, a simulator could generate data about system incidents/accidents in different operating and environmental conditions.

The database is used to build the structure of the risk model, presented in the second layer in the figure. Real-time data can be utilized to update the input parameters of the risk model—in real time [3]. Sensors and actuators may convey useful information about the system and environmental conditions in real time [4]. In long-term operation, the real-time data should be collected and stored in a database and treated

T. Parhizkar et al., *Online Probabilistic Risk Assessment of Complex Marine Systems*, Springer Series in Reliability Engineering,
https://doi.org/10.1007/978-3-030-88098-9_3

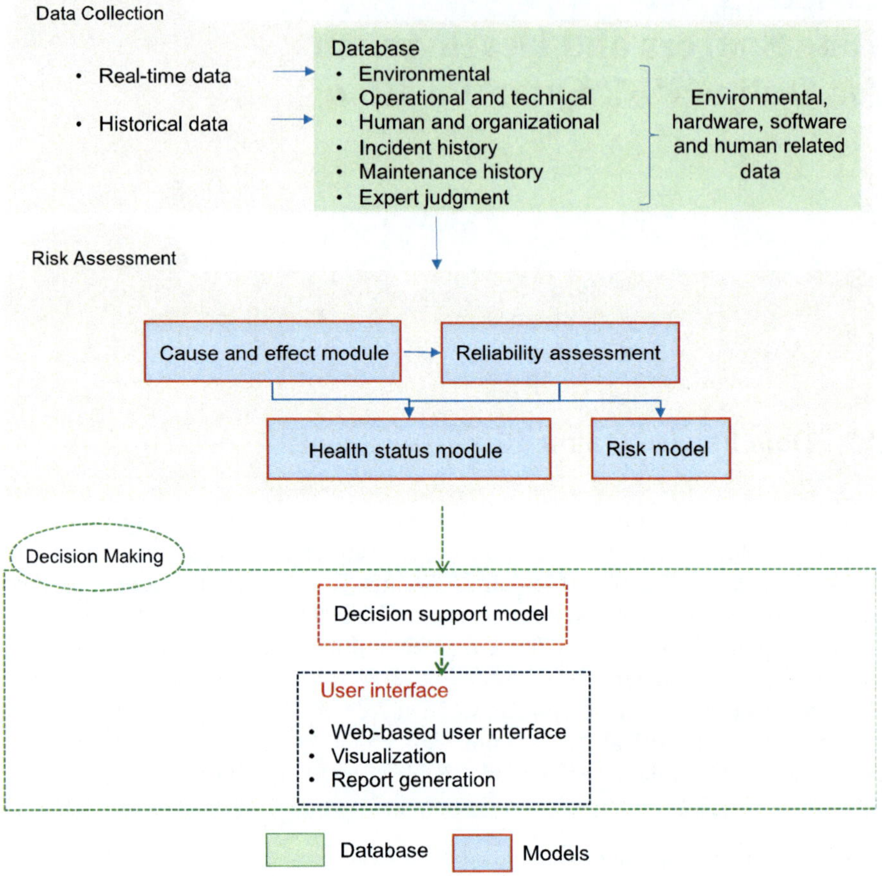

Fig. 3.1 Data flow diagram of online risk assessment and decision support module. Since decision support is outside the scope of the book, this part is shown in white

as the historical data. These data could be used to update failure data distributions and the structure of the risk model in specific intervals determined by users.

The risk assessment layer consists here of four main modules, i.e., the cause-and-effect module, reliability assessment module, risk model, and health status module; the flow of data in these modules is presented in Figs. 3.2, 3.3, 3.4, 3.5 and 3.6, respectively. In all figures, historical data is taken as inputs, but the type of historical data could be different. Data-driven models, physic-based models, or a hybrid of data-driven and physic-based model can also provide input in the data flow and the modules (Sect. 3.3).

The outputs of the risk assessment are used in the next layer; decision-making, to assist operators to make better decisions. In this layer, in addition to risk assessment results, other system performance indices may be considered, and the best decision

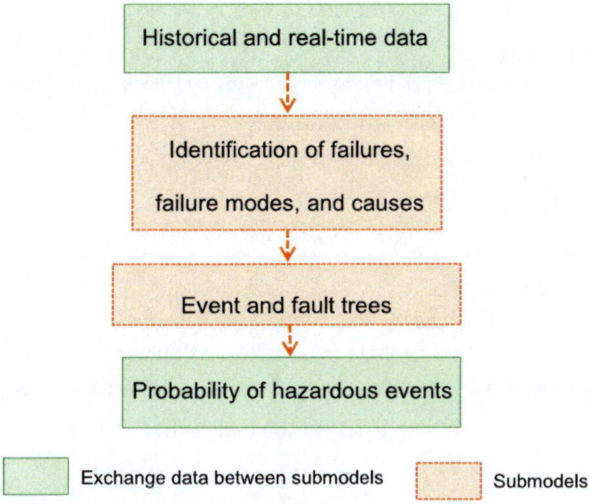

Fig. 3.2 Flow diagram of cause-and-effect module

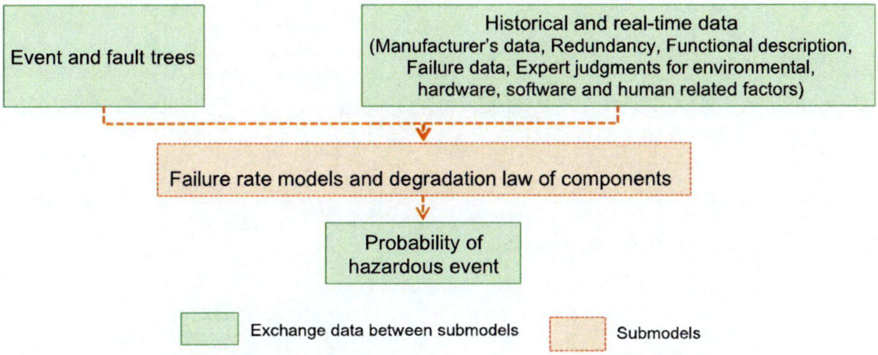

Fig. 3.3 Data flow diagram of reliability assessment module

scenario could be obtained and visualized. Since decision support is outside the scope of the book, this part is shown in white.

The first module in the risk assessment layer is the cause-and-effect module, which purpose is to develop failure data that can be used in the reliability assessment module. In this module, failures are identified based on historical and real-time data. Then the failure modes, causes, and mechanisms should be studied, as a basis for developing, for example, event and fault trees. These trees help identify, structure, and analyze causes and consequences of failures and risk-influencing factors (See Chap. 11. Glossary), [5, 6].

The next module in the risk assessment layer is the reliability assessment module. Figure 3.1 illustrates the flow of data in this module. As can be seen, event and fault trees (derived from cause-and-effect module) and system historical and real-time data

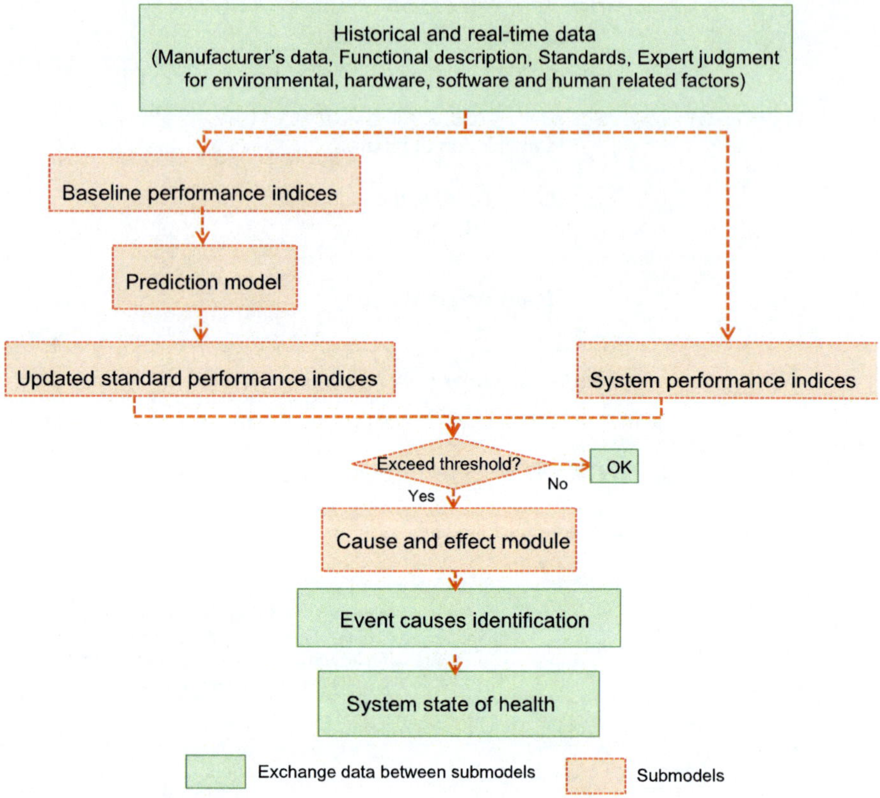

Fig. 3.4 Flow diagram of health status module

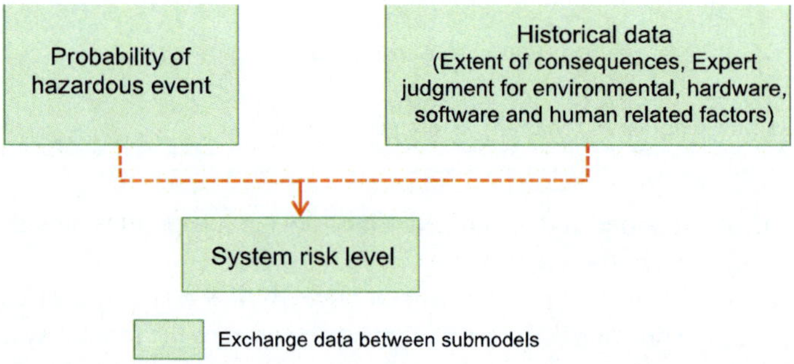

Fig. 3.5 Flow diagram of risk model

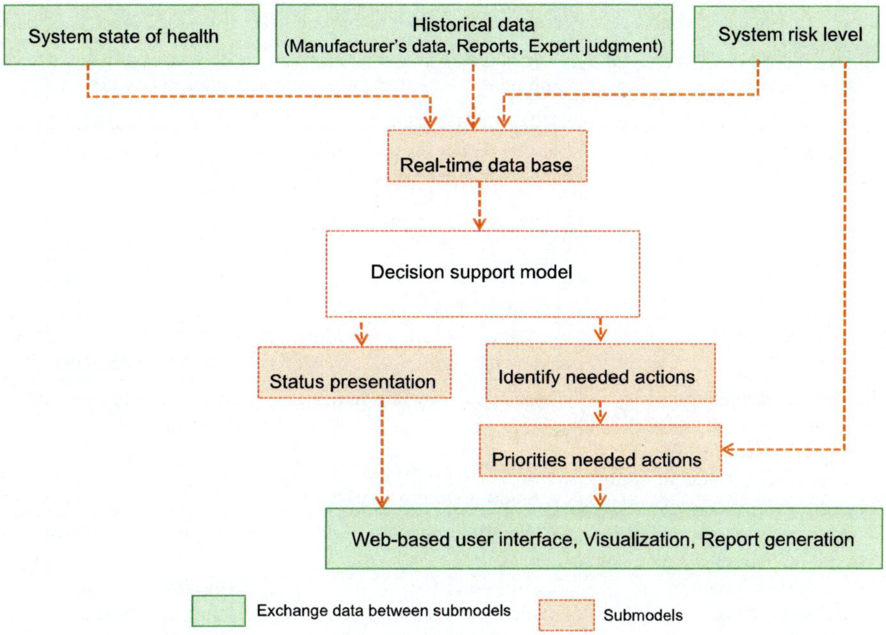

Fig. 3.6 Flow diagram of decision support model

are taken as inputs. In this module, failure probabilities are calculated using failure rate models and degradation laws of components. According to the governing failure mechanisms in a system, a failure model and degradation laws should be developed. For instance, degradation principal equation in a mechanical system is completely different from a system that performs electrochemical reactions [7–10]. Table 3.1 presents different failure rate models that could be used in complex systems.

The failure rates of components can, for example, be used in fault trees to estimate basic event probabilities of the system [12].

The purpose of the system health status module is to provide information about the current condition of the system. The flow diagram of the health status module is depicted in Fig. 3.4. In such modules, historical and real-time data and/or system simulations are used to develop baselines of the systems' performance. The performance baselines are represented through indicators of system health status in a standard operating condition. The real-time data updates the performance indices, according to the system operating conditions. In the real time, the baseline is compared with the updated data from the system. If the difference between the system performance indices exceeds the acceptable threshold, the system has a health issue. In this case, the cause-and-effect module is run and analyzed, and causes to the performance deviations can be identified. The system health status is OK if the system performance indices are within an acceptable range [13].

Table 3.1 Typical probability (or frequency) models in probabilistic risk assessment and their parameters [11]

Basic event type	Commonly used models of basic event probability	Data required to quantify models
Initiating event	Poisson model for probability of seeking k events in time t: $\Pr(k) = e^{-\lambda t} \frac{(\lambda t)^k}{k!}$ where t: Mission time λ: Frequency	Number of events k in time t
Component fails on demand	Constant probability of failure on demand, or q	Number of failure events k in total number of demands N
Standby component fails in time, or component changes state between tests (failures revealed on functional test only)	Constant standby failure rate $Q = 1 - \frac{1-e^{\lambda_s T_s}}{\lambda_s T_s}$ T_s: Time between tests λ_s: Standby failure rate	Number of events k in total time in standby T
Component in operation fails to run, or component changes state during mission (state of component continuously monitored)	Constant failure rate $U = 1 - e^{\lambda_0 T_m} \approx \lambda_0 T_m$ T_m: Mission time λ_0: Operating failure rate approximation is adequate when $\lambda_0 T_m < < 1$	Number of events k in total exposure time T (total time standby component is operating, or time the component is online)
Component unavailable due to test	$Q = \frac{T_{\text{TD}}}{T_s}$ T_{TD}: Test duration (only in the case of no override signal) T_s: Time between tests	Average test duration (T_{TD}) and time between tests (T_{s})
Component unavailable due to corrective maintenance (failure revealed only at periodic test, or preventive maintenance performed at regular intervals)	$Q = \frac{T_{\text{U}}}{T_T}$ T_U: Total time unavailable while in maintenance (out of service) T_T: Total operating time	Total time out of service due to maintenance actions while system is operational, T_U, and total operating time T_T
Components unavailable due to unscheduled maintenance (Continuously monitored components)	$Q = \frac{\mu T_R}{T_T}$ T_R: Average time of a maintenance outage ["repair time"] μ: Maintenance rate	Number of maintenance actions r in time T (to estimate μ)
Standby component that is never tested. Assumed constant failure rate	$Q = 1 - e^{-\lambda_m T_p}$ T_P: Exposure time to failure λ_m: Standby failure rate	Number of failures r, in T units of (standby) time
Common-cause failure probability	α_1 through α_m, where m is the redundancy level	n_1 through n_m where n_k is the number of CCF events involving k components

Note Data needed to estimate those parameters is listed in the right-hand column. Other model parameters (such as "mission time") are determined by the application

The calculated failure probabilities from the reliability assessment module, in addition to failure consequences, are used in the risk model to evaluate the risk level of the system [14]. Risk here is the probability of an incident occurrence multiplied with the consequences resulting from the incident. The flow diagram of this process is illustrated in Fig. 3.5.

In the last layer in the data flow diagram, the output of the risk assessment layer can be used to perform the evaluations for decision support. The system health status, risk level, and other required historical data should be stored in a database to update the risk, health, and reliability modules in operation (Fig. 3.6).[1]

3.2 Data Collection and Preparation

In order to perform dynamic risk assessment of a complex system, the boundary of the system under study should be defined clearly. Then, as presented in Fig. 3.1, the first step of probabilistic risk assessment (PRA) is data collection. One of the data collection methods is reviewing incident reports and failure records of the complex system if these are available [14]. For instance, in dynamic positioning (DP) systems, the international marine contractor's association (IMCA) provides information about DP systems events and incidents [15]. IMCA is the international trade association representing offshore, marine, and underwater engineering companies. The IMCA DP station keeping has been publishing an annual incident report for DP systems from 2000. Each report contains an overview and analysis of multiple incidents. Figure 3.7 presents an example of an IMCA incident report from 2004, [16].

It should be noted that it is difficult to verify the quality of the incidents included in the IMCA reports, since they are not in their original form. Also, it may be that the original incident investigation was not thorough enough to shed light on the causal factors or that it was not performed by people with the right expertise. Often the description of the incidents in the IMCA reports is very brief and lacking in detail, which leaves room for interpretation and subjectivity. These are some of the limitations that impact data quality. In addition, underreporting has to be assumed. The IMCA reports are based on incidents that have been shared by various vessel owners voluntarily; hence, it can be assumed that it is not a complete overview, but how many incidents are missing from the IMCA reports can only be guessed. It might, for example, be that there is significant underreporting of incidents caused by human error due to a possible shame and blame culture. The authors believe that despite the limitations in being able to verify the data quality from the incident reports, the information utilized in this book provides a good starting point for the development of a risk model, which can be updated and refined at a later stage when more information becomes available.

After gathering available data on incidents, the data should be analyzed to extract useful information. For instance, in Fig. 3.7, the top level of the flow diagram presents

[1] Human–machine interface and decision-making are topics outside the scope of this book.

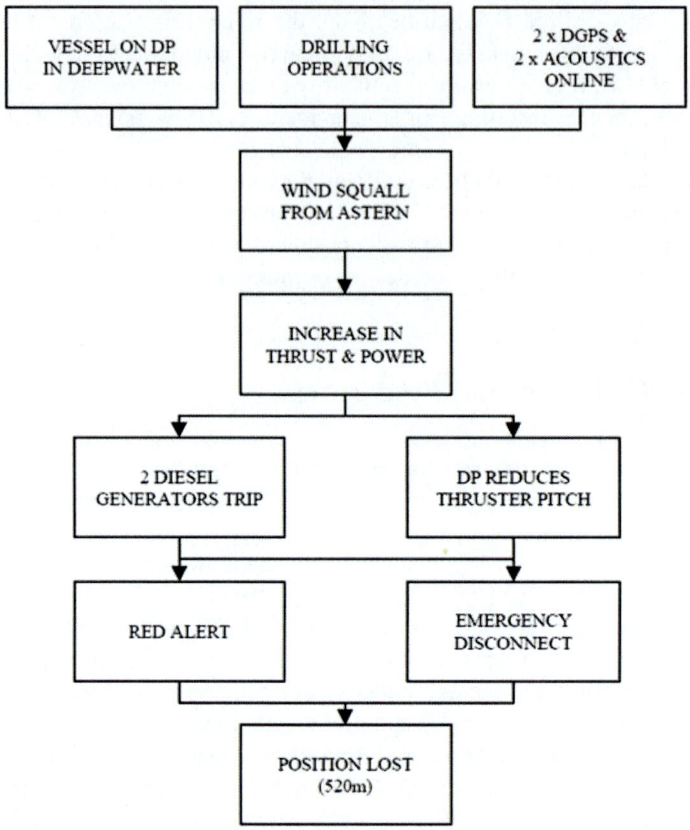

Fig. 3.7 An incident report from IMCA annual dynamic positioning (DP) report [16], (DGPS: differential global positioning system)

system and environmental characteristics. From the figure, it could be inferred that a DP drilling unit was operating, and two differential Global Positioning System (DGPS) and two acoustic reference systems were online at the time of incident. So, the type of DP system (such as DP drilling, supply, diving, flotel, cable/pipe laying) and some system characteristic could be identified from the first layer. The following boxes in the figure present the events sequentially until the incident happens. The end states of the incidents (consequences) are categorized into drift-off, drive-off, loss of position, and potential incident. Drift-off is an undesired event that occurs when the DP system cannot counteract environmental forces due to, for example, power loss. Drive-off is an event that occurs if the DP system actively drives the vessel away from the desired position and/or heading using thruster force, [17]. Loss of position is considered as a situation that vessel losses its position, but the reason (drift-off or drive-off) is not mentioned in the incident report clearly. Potential incidents are events that could have led to loss of position. In the incident presented in Fig. 3.7, loss of position is the end state of the incident.

Figures 3.8, 3.9, 3.10 and 3.11 present the results from the analysis of more than 700 incidents reported from 2004 to 2015, [18]. The incidents are distributed based on type of DP vessel and operation (Fig. 3.8), operating phase (Fig. 3.9), incident trigger (Fig. 3.10), and consequences (Fig. 3.11). The "unknown" sections in the figures represent the fraction of incidents for which no information was available in the incident reports.

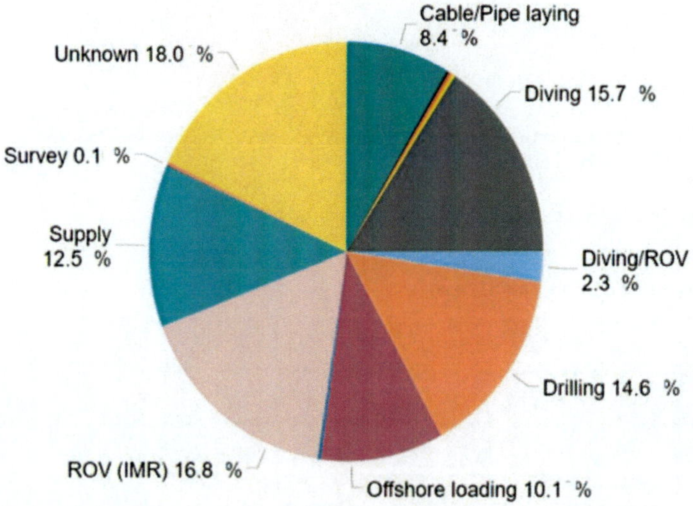

Fig. 3.8 Incidents distributed by DP vessel types, (ROV: remotely operated vehicle, IMR: inspection, maintenance, and repair)

Fig. 3.9 Incidents categorized by DP operating phase

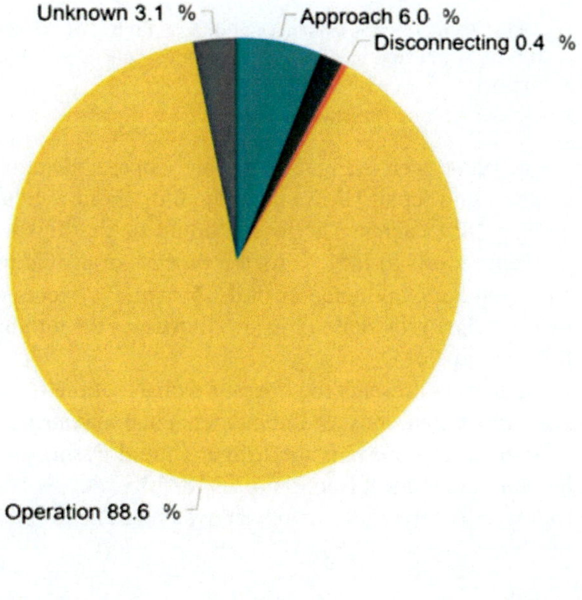

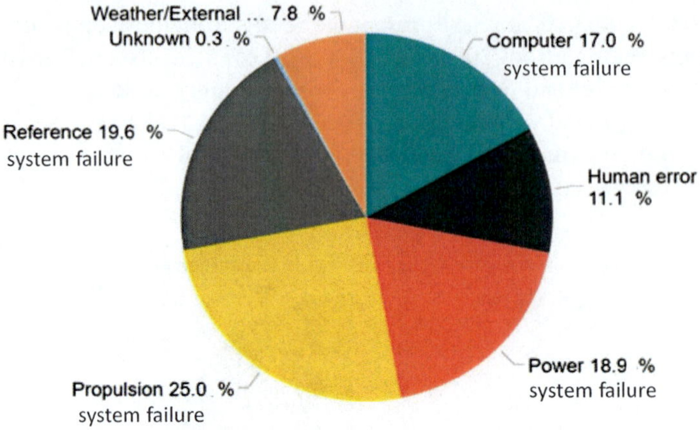

Fig. 3.10 Incidents triggers (initial events) distribution

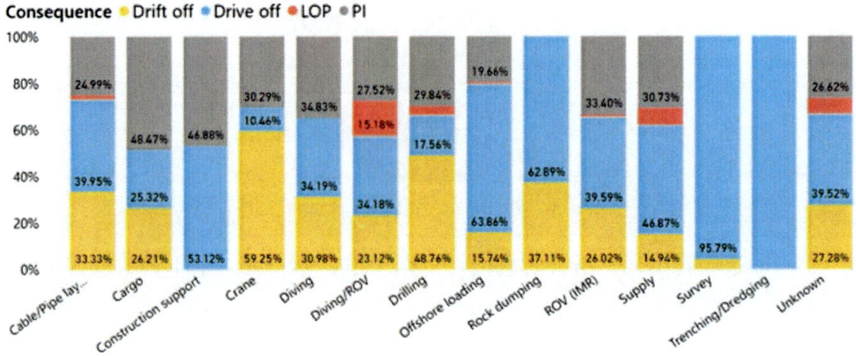

Fig. 3.11 Incident's consequences of different types of DP vessels and operations, (LOP: loss of position, PI: potential incident, ROV: remotely operated vehicle, IMR: inspection, maintenance, and repair)

As can be seen in Fig. 3.8, remotely operated vehicles (ROVs), inspection, maintenance, and repair (IMR) have the most failure incidents, followed by diving and drilling DP systems. The reason could be the higher number of vessels providing incident reports to IMCA. As the process of providing incident reports to IMCA is not mandatory, all incidents of different DP systems would not be recorded. Therefore, the data provided in Fig. 3.8 illustrates the number of incidents that is recorded for each type of DP system.

Figure 3.9 presents the operating phase of the DP system at the time of incident. As is illustrated, most of the incidents happen during operation time.

Figure 3.10 presents the trigger (initial event) distribution of incidents. As can be seen, the highest fraction of incidents is due to failure in the propulsion system, followed by reference and power systems. The propulsion system includes thrusters and all the control system and auxiliary systems required for thrusters to work properly.

Figure 3.11 presents the incident consequences in different types of DP vessels and operations. As is shown, the proportion of consequences varies in different DP systems. However, in all DP systems, loss of position (LOP) has the least frequency.

The information above may help in the development of the DP system fault trees based on the presented cause-and-effect module in Fig. 3.2. The data presented in Fig. 3.10 helps to identify the main causes of failure, and by reviewing the reports (Fig. 3.7) in more detail, the root cause of failure of each component can be derived. In addition, Fig. 3.11 presents the consequences of the incidents that may help us in developing event sequence diagrams or event trees.

3.3 Quantifying the Probability or Frequency of Basic Events

As presented in the reliability assessment module (Fig. 3.3) events' failure probabilities or frequencies are a function of system technical operating conditions, environmental and operators' characteristics. In order to quantify failure probabilities, failure models and degradation laws of the system components should be developed. These could be quantified based on physics-based, data-driven, and hybrid methods. In physics-based methods, physics and chemical governing equations are utilized to find the relation between input data and outputs. Such methods require comprehensive knowledge about system behavior and components interactions, e.g., modeling a vessel based on the hydrodynamic principles. The outputs may, for example, be represented in a digital twin of the system.

Data-driven methods use different machine learning methods and generate a predictive function accordingly. Such methods require a huge amount of data regarding normal operation, deviations from "normal", and different operating and environmental conditions. In the lack of data, the developed models using such methods would only provide acceptable results in a specific range of operating and environmental conditions. For instance, the failure of space shuttle systems as a function of environmental conditions could not be modeled purely based on the data-driven methods as there is not enough data on its critical failure modes in different environmental conditions [19, 20].

In recent years, hybrid methods have been proposed that try to utilize the advantages of both physics-based and data-driven methods. The hybrid methods take the advantage of physical and chemical governing equations along with available data to predict system behavior with a minimum uncertainty. There are multiple studies that have developed hybrid methods to model different complex system behaviors [21, 22]. These studies use different methodology according to the available knowledge and data about the complex system. In the example DP system, failure probabilities of technical components are evaluated using hybrid methods, as presented in [11].

Table 3.2 Drilling dynamic positioning (DP) operational time over 2004 to 2015 in worldwide DP drilling [24]

Year 20XX	04	05	06	07	08	09	10	11	12	13	14	15
Operating year	49	49	50	50	72	80	89	87	93	98	104	110

Table 3.1 summarizes underlying probability models that may be relevant to risk assessment [11].

Table 3.1 also indicates the required data to estimate various parameters. The data type required varies depending on the event type, and on how its frequency or probability is modeled. For example, the probability of "components fail on demand" could be quantified based on the number of failure events, and exposure or success data (total operating time). In the DP example, discussed in Sect. 3.2, the number of incidents is available. According to Table 3.1, the operating hour of DPs is another type of data that is required for initial event frequencies quantification. Based on the review performed, the operational time of DP drilling vessel from year 2004 to 2010 is gathered from drilling semi-submersibles and drill ships worldwide [24], and the operating hours for the remaining years are estimated based on the available trend of the data, presented in Table 3.2. The sum over 12 years is 937 years of operation.

The failure frequencies of all initial events of the thruster component in DP drilling units are presented in Table 3.3. Component's failure and causes to the failures are derived for the DP drilling unit from 117 IMCA incident reports from 2004 to 2015 that were available online [18]. Most of the incident reports present the components involved in the incident. For instance, the incident reported in Fig. 3.7 indicates that the diesel generator trip is the reason for the LOP [16]. Failure frequencies of other components of the DP system, i.e., reference, control, and power systems, are determined and quantified in [25].

There are other studies that have focused on failure frequency estimation of maritime systems, such as [26, 27]. The level of detail is different in these studies, for instance, in [27] failure rates of main components, such as the reference system and main engine are estimated. In [27], failure rates of the components of the main engine, such as medium voltage (MV) power bus, medium voltage circuit breaker, and power management system are presented.

Failure probabilities are the outputs of the reliability assessment module, presented in Fig. 3.3. These values and failure consequences enter the risk model, and the system risk level will be calculated, as presented in Fig. 3.5. The failure consequences depend on the failure mode, DP system type, operation mode, and environmental conditions that could be quantified by experts.

Table 3.3 Failure frequency of thruster subsystems in dynamic positioning drilling unit based on international marine contractor's association (IMCA) incident reporting [25]

Components failure	Causes	Frequency per 10^6 h[1]
Thruster unit and drive	Main coupling error	12.2
	Direct current (DC) motor field problem	12.2
	Unknown	48.8
Control system	Control unit's programmable logic controller (PLC) error	24.4
	Fuses, relays, printed circuit board (PCB), signal amplifiers error	24.4
	Outstation internal power distribution error	24.4
	Unknown	60.9
Feedback signal	Loose or broken linkages	12.2
	Faulty potentiometers	12.2
Hydraulics (controllable pitch propeller, clutch)	Control valves, proportional, solenoids, switches error	12.2
	Low hydraulic oil pressure/pitch pump fault	12.2
Cooling, lubrication, air, ventilation	Thruster brake fault	12.2
	Oil leaks	12.2

[1]Numbers are rounded off for the purpose of reporting

References

1. Parhizkar T, Mosleh A, Roshandel R (2017) Aging based optimal scheduling framework for power plants using equivalent operating hour approach. Appl Energy 205:1345–1363
2. Parhizkar T, Rafieipour E, Parhizkar A (2021) Evaluation and improvement of energy consumption prediction models using principal component analysis based feature reduction. J Cleaner Prod 279:123866
3. Hogenboom S, Parhizkar T, Vinnem JE (2021) Temporal decision-making factors in risk analyses of dynamic positioning operations. Reliab Eng System Saf 207:107347
4. Parhizkar T, Balali S, Mosleh A (2018) An entropy based bayesian network framework for system health monitoring. Entropy 20(6):416
5. Parhizkar T, Aramoun F, Saboohi Y (2020) Efficient health monitoring of buildings using failure modes and effects analysis case study: air handling unit system. J Build Eng 29:101113
6. Parhizkar T, Aramoun F, Esbati S, Saboohi Y (2019) Efficient performance monitoring of building central heating system using Bayesian Network method. J Build Eng 26:100835
7. Parhizkar T, Hafeznezami S (2018) Degradation based operational optimization model to improve the productivity of energy systems, case study: Solid oxide fuel cell stacks. Energy Convers Manage 158:81–91
8. Parhizkar T, Roshandel R (2017) Long term performance degradation analysis and optimization of anode supported solid oxide fuel cell stacks. Energy Convers Manage 133:20–30
9. Parhizkar T (2018) Long-term degradation-based modeling and optimization framework. In Handbook of research on predictive modeling and optimization methods in science and engineering (pp 192–220). IGI Global

10. Roshandel R, Parhizkar T (2016) Degradation based optimization framework for long term applications of energy systems, case study: solid oxide fuel cell stacks. Energy 107:172–181
11. Stamatelatos M, Dezfuli H, Apostolakis G, Everline C, Guarro S, Mathias D, Mosleh A, Paulos T, Riha D, Smith C, Vesely W, Youngblood R (2011) Probabilistic risk assessment procedures guide for NASA managers and practitioners
12. Parhizkar T, Utne IB, Vinnem JE, Mosleh A (2021) Dynamic probabilistic risk assessment of decision-making in emergencies for complex systems, case study: dynamic positioning drilling unit. Ocean Eng 237:109653
13. Roemer MJ, Nwadiogbu EO, Bloor G (2001) Development of diagnostic and prognostic technologies for aerospace health management applications. In 2001 IEEE aerospace conference proceedings (Cat. No. 01TH8542), vol 6. IEEE, pp 3139–3147
14. Parhizkar T, Utne IB, Vinnem JE, Mosleh A (2021) Supervised dynamic probabilistic risk assessment of complex systems, part 2: application to risk-informed decision making, practice and results. Reliab Eng Syst Saf 208:107392
15. Dynamic positioning system incident reports, International Marine Contractors Association, https://www.imca-int.com/dp-events/, 17 June 2021
16. The International Marine Contractors Association (2005) Dynamic positioning station keeping review, incidents and events reported for 2004. International Marine Contractors Association
17. Rokseth B (2018) Safety and verification of advanced maritime vessels—an approach based on systems theory. PhD. Dissertation. Department of Marine Technology, Norwegian University of Science and Technology
18. The International Marine Contractors Association (2005–2017) Dynamic positioning station keeping review, incidents and events reported for (2004–2015). International Marine Contractors Association
19. Hu Y, Parhizkar T, Mosleh A (2021) Guided simulation for dynamic probabilistic risk assessment of complex systems: concept, method, and application. Reliab Eng Syst Saf. p 108047
20. Nejad HS, Parhizkar T, Mosleh A (2021) Simulation based probabilistic risk assessment (simpra): risk based design. 31st European safety and reliability conference. 19–23 September 2021, Angers, France
21. Parhizkar T, Mosleh A (2022) Guided probabilistic simulation of complex systems toward rare and extreme events. In the 68th Annual reliability & maintainability symposium (RAMS), Tucson, Arizona
22. Parhizkar T, Vinnem JE, Utne IB, Mosleh A (2020) Supervised dynamic probabilistic risk assessment of complex systems, part 1: general overview. Reliab Eng Syst Saf, p 107406
23. Baraldi P, Cadini F, Mangili F, Zio E (2013) Model-based and data-driven prognostics under different available information. Probab Eng Mech 32:66–79
24. Chen H, Nygärd B (2016) Quantified risk analysis of DP operations-principles and challenges. In SPE international conference and exhibition on health, safety, security, environment, and social responsibility. Society of Petroleum Engineers
25. Parhizkar T, Hogenboom S, Vinnem JE, Utne IB (2020) Data driven approach to risk management and decision support for dynamic positioning systems. Reliab Eng Syst Saf 201:106964
26. Vedachalam N, Ramadass GA (2017) Reliability assessment of multi-megawatt capacity offshore dynamic positioning systems. Appl Ocean Res 63:251–261
27. Soares CG, Santos TA (2018) Progress in maritime technology and engineering: Proceedings of the 4th international conference on maritime technology and engineering (MARTECH 2018), 7–9 May 2018, Lisbon, Portugal

Chapter 4
Human, Hardware, and Software Interactions in Risk Assessment

4.1 Risk of Human, Hardware, and Software Failures

Hardware reliability is not the only key to the effectiveness of online risk assessment in maritime systems; there is also a need to understand and consider the influences on risk from human and organizational factors [1, 2], and software failures [3, 4], as well as their interactions and cascading effects.

There have been studies of specific challenges of considering human and organizational factors in risk assessment [5–7]. Human reliability analysis (HRA) aims to quantify the probability of human error for a given task. In [8], accident investigation and QRA are combined using available information from a precursor incident as input to QRA-methodology. This method includes an extended understanding of human and organizational factors in accidents and their prevention. The result helps identifying hazards, probabilities, safety barriers, and possible consequences and gives a more complete cause and risk picture in complex systems.

Gran et al. [9] proposes a risk model that integrates organizational, human, and technical factors to develop more representative models for calculation of leak frequencies as a function of the volume of manual operations and interventions. Vinnem et al. [10] proposes a quantitative risk analysis model that considers operational barriers in event trees and fault trees, as well as risk-influencing factors that determine the basic event probabilities in the fault trees. Human error probabilities, common cause modeling, and their interactions are discussed. The proposed model in this study could be used to demonstrate the importance and effects of improving human and organizational aspects. Petro-HRA is an HRA method adapted for offshore oil and gas industries. In Petro-HRA [11], human–machine interaction is considered using performance shaping factors (PSFs). PSFs are factors that could enhance or degrade human performance. In Petro-HRA, the effect of human performance on machine functionality is modeled using PSFs.

In [12], the causes and barrier failures of nine reported accidents and incidents with DP systems occurring over a 16-year period (2000–2015) in the North Sea

T. Parhizkar et al., *Online Probabilistic Risk Assessment of Complex Marine Systems*, Springer Series in Reliability Engineering,
https://doi.org/10.1007/978-3-030-88098-9_4

were studied. The results show that the event sequence of each accident and incident together with their immediate causes, root causes, as well as failed or missing barriers, includes human, organization, and technical aspects. Results indicate that human errors related to the designer, operators, maintenance personnel, etc., and failure of various organizations, such as third-party verification, vendors, authorities, etc. are involved in the nine reported incidents and accidents. This emphasizes that all the parties involved have obvious responsibilities for the safety management of maritime systems.

There are other studies that have indicated the importance of considering human, organization, and technology interactions in risk assessment of maritime systems [13–16]. A better understanding of how different human, organizational, and technological elements interact could explain how different failure scenarios in a complex system may occur [17].

Software failures and impact on system risk are less understood. Whereas risk and reliability assessments for hardware failures and human errors are relatively mature, the same is not true for software [3, 4]. It is important to address software failures along with any resulting failure modes in the system [18]. A challenge is that software failures are different from hardware and human failures. In general, they are not considered stochastic and appear depending on the value of system parameters. Software reliability often focuses on the number of bugs remaining in a code but is not preoccupied with the potential impact on the system as a whole [19]. With regular upgrades, the past performance of software cannot be used as an indication of its future performance, which is particularly challenging with autonomous control systems with self-learning abilities.

A challenge with complex systems in the marine and maritime domain is that they consist of several different components and subsystems, often from many different vendors. This means that even if the components function reliably individually, new system-level properties arise when the components are integrated into a larger system. Achieving a sufficient understanding of how complex software intensive systems may fail is very difficult [20]. In the maritime industry, for example, it is often assumed in reliability and redundancy analyses (FMEA) of the DP system that the software does not fail. This critically limits the understanding and the assessment of potential hazards.

4.2 Risk Assessment of Software

One of the most important parts of automated maritime systems is software. Software is involved in different components of the system including sensors, monitoring systems, control systems, navigation systems, and human–machine interfaces.

Software does not age, and its behavior is not dependent on the environmental and operating conditions of the system. With this point of view, software error is deterministic. There may, however, be unknown errors that are not detected through testing and verification of software. These unknown errors can be triggered in different ways and have a probability of occurrence.

Another type of probabilistic software failure is caused by the mismatch between software requirements and unpredicted environmental and operational conditions. Each software has a specification document that presents all the requirements to the software operation. However, in real applications, some situations can happen which are outside the operational envelope (defined in the specification) of the software. In these cases, a software failure could happen. As the environmental and operational conditions of a system are uncertain, the software failure can be uncertain and have a probabilistic behavior. The third type of probabilistic software failure is because of hardware degradation. Each software resides on one or several hardware components. The hardware components degrade as time passes. The degradation of components depends on multiple uncertain factors and may lead to probabilistic software failures.

Different models are proposed to evaluate software risk and reliability. One method is FMEA. In [21–23], a failure mode taxonomy for software is presented and discussed based on the FMEA method. Another method, which is widely used, is the dynamic flow graph methodology (DFM). This method is a directed graph with functional relations that performs analyses at discrete time intervals. Al-Dabbagh [24] has developed a network control system based on DFM. The US National Aeronautics and Space Administration (NASA) proposed a context-based software risk assessment methodology (CSRM) based on the DFM. This method is developed to identify, analyze, and mitigate risks associated with space missions [3, 25].

Software failures are considered in dynamic probabilistic risk assessment of complex systems. For instance, [26] and [27] consider software behavior in dynamic fault trees of the system, and the risk level is evaluated accordingly. In [28], software risk is considered as a module in the dynamic risk assessment model of the system. Software failure modes are reviewed and categorized in [4].

Herein, a guided software failure scenario generation framework is proposed (Fig. 4.1). The framework aids in the identification of software-related failure scenarios in a complex system.

As can be seen in Fig. 4.1, the guided failure scenario generation could be broken down into six steps, which are explained in the following subsections.

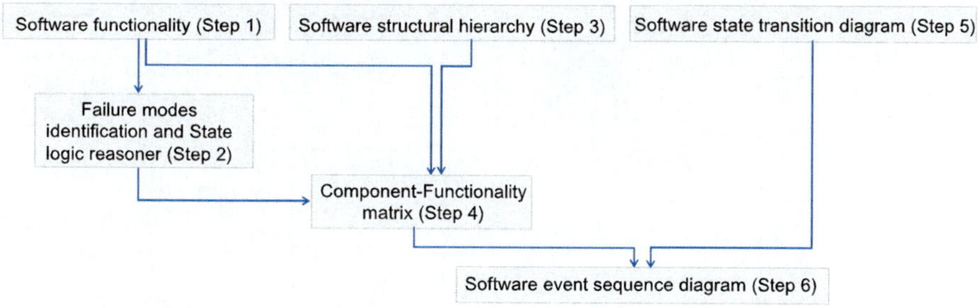

Fig. 4.1 A framework for guided failure scenario generation of software

4.2.1 Step 1: Defining Software Functionalities

In this step, the main functions of the software in a complex system should be identified. On the hardware side, functions of each components are comprehensively developed. For instance, the function of a pump is to increase pressure of the fluid, or a heat exchanger transfers heat from a hot stream to a cold stream. Defining the main functions of software, however, may be complicated. Software functionalities should be defined based on the software application and the goal of software operation, e.g., the main function of a software installed on a unit converter is unit conversion; however, the main function of a software on a valve controller is to control (open, close) a valve. According to [29], "control," "configure," and "measure" could be identified as the main functions of software. Figure 4.2 presents a general functional tree for software based on [29].

"Configure" as a function has two main purposes; configuration, i.e., parametrization of a component, and the conversion of a signal, to improve its quality or make it suitable for further use by other components. Control refers to the function of executing actions that will change the state if the system, i.e., act on the process variables of the system. This includes opening, closing, regulating physical flow, changing operational modes, turning components on or off, etc. Measure refers to a function collecting information of the environment and on the system that is processed to be used for configuration or control. Figure 4.2 is a general software functional tree, and the analyst should develop the software functional tree for the system under study based on the software application and the goal of software operation.

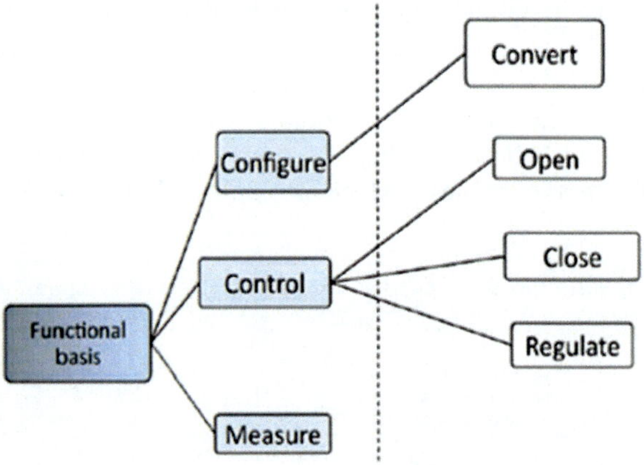

Fig. 4.2 A software functional tree [29]

4.2.2 *Step 2: Developing State Logic Reasoner*

It is very difficult to identify a functional failure or a failure mode of a software due to the unknown location of the failure, limited knowledge, and complexity of the software behavior. It may also be discussed whether the term "failure mode" from hardware reliability actually is relevant to use for software since there is no degradation as such, and if "failure event" would be more applicable. In this book, however, we use the term failure mode for software failures, for the sake of simplicity.

One common method for software failure identification is based on the flow state logic reasoner [29]. The flow state logic reasoner is a method for reasoning on the state of data flows. This allows for the assessment of failure propagation over potential data flows that were not considered in a functional representation of a nominally functioning design.

To capture the possibility of failure propagation of the potential data flows, the flow state logic reasoner identifies the state of any data flow in the system of interest for any given system state. Flow states can be separated as one normal state representing the designed data flow and four states representing the failure modes. The different types of failure modes can be categorized as [3]:

1. Process function failure modes
2. Time-related failure modes
3. Value-related failure modes
4. Interaction failure mode.

Thieme et al. [4] represents a generic functional decomposition of a software that includes inputs and outputs, and with outputs being used as input to other functions. The elements of the function in Fig. 4.2 may have different functional failures or failure modes. The process section is where the functional behavior and computation are executed, turning input(s) to output(s). A function in principle has at least one output. This maybe a numerical value or a function call to other functions. Input is the output of another function or is provided by an external interface. An output of a function can be input to several other functions. Each function can have several inputs and outputs with a specific data type and range (Fig. 4.3).

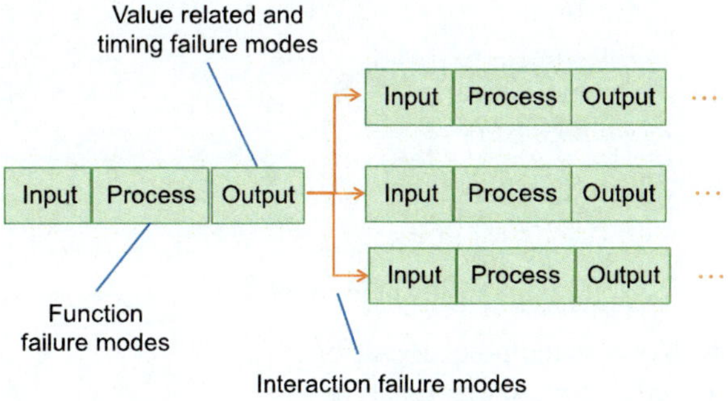

Fig. 4.3 Functional decomposition of software

The input, process, and output could have different failure modes. The input could have interaction failure modes, the process can fail because of functional failure modes, and the output could fail due to value and time-related failure modes. In the following, some generic examples of these types of failure modes are presented [3].

The function failure modes include, but is not limited to:

- Omission of a function/missing operation
- Incorrect functionality
- Additional functionality
- No voting
- Incorrect voting
- Failure in failure handling.

The time-related failure modes include, but is not limited to:

- Output provided: too early
- Output provided: too late
- Output provided: spuriously
- Output provided: out of sequence
- Output provided: not in time
- Output rate
  - Too fast
  - Too slow
  - Inconsistent
  - Desynchronized
- Duration
  - Too short
  - Too long
- Recurrent functions scheduled incorrectly

The value-related failure modes include, but is not limited to:

- No value
- Incorrect value
  - Too high
  - Too low
  - Opposite/inverse value
  - Value is zero
- Value out of range
  - Datatype allowable range
  - Application allowable range
- Noisy value/precision error
- Value with wrong datatype

- Elements in a data array/structure
  - Too many
  - Too few
  - Data in wrong order
  - Data in reversed order
  - Enumerated value incorrect
- Non-numerical value
  - NaN
  - Infinite
- Redundant/frozen value
- Correct value is validated as incorrect
- Incorrect value is validated as correct
- Data is not validated.

The interaction failure modes include, but is not limited to:

- Diverted/incorrect functional call
- No call of next functionNo priority for concurrent functions
- Incorrect priority for concurrent functions
- Communication protocol-dependent
- Unexpected interaction with input/output (IO) boards
  - File/database wrong name
  - File/database invalid name/extension
  - File/database does not existFile/database is open
  - Wrong/invalid file format
  - File head contains error
  - File ending contains error
  - Wrong file length
  - File/database is empty
  - Wrong file/database contents.

In this step, we should develop different logic rules according to the software design specification document[1] to identify system status (success[2] or failure) and potential failure modes. Figure 4.4 presents a flow state for a very simple software function which takes average of inputs (A and B) and provides the result (C).

In the presented example Fig. 4.4, we should present the logic rules for input, process, and outputs based on the design specification. For instance, if the output is assumed to be a number, we have the following logic rule (Fig. 4.5).

[1] Software design specification is the first step of software development. In this step, all specifications, requirements, and standards that should be followed and implemented for coding, run-time, test, updating, and maintenance will be specified and documented.

[2] Success refers to safe mode.

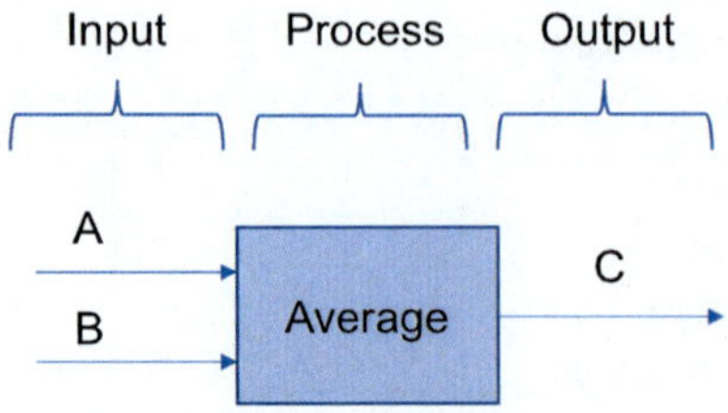

Fig. 4.4 Flow state of a software with one function (taking average)

If C is a number

Then, Output value-related function=OK

Else

Output value-related function=Fail-Value is out of range

Fig. 4.5 An example of a logic rule

We should develop all other logic rules for input, process, and output, at this step. Using these logical rules, we can identify system state and corresponding software failure modes.

4.2.3 Step 3: Defining Software Structural Hierarchy

The software structural hierarchy helps identifying all hardware components involved in software operation and functionality. The structural hierarchy will be used in Step 6 to generate software event sequence diagrams. In general, a structural hierarchy refers to the subsystems and components that the system is built of. The "AND" relation shows the distinct elements while "OR" is used to present the redundancies.

Figure 4.6 presents the structural hierarchy of the example software. The main subsystems are configuration, controller, sensor, data logging systems, and programming codes. These are all the hardware components required for the software to reside and operate. For instance, the configuration function running on a configuration component. As is presented in the figure, the configuration subsystem has a unit converter component. The controller system has shutdown, isolation, and regulation controller components. A sensor could be multiple components, such as pressure and temperate sensors. The presented structural hierarchy could be adjusted for different types of software based on the system boundary and software application.

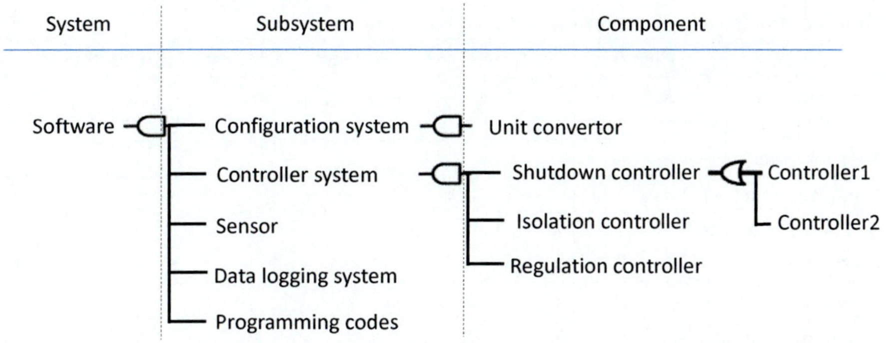

Fig. 4.6 Structural hierarchy of example software system

4.2.4 Step 4: Developing Component-Functionality Matrix

In this step, the relationships between software components and software functions should be mapped. This is important in order to understand which components are involved in the functions and failure modes, as this may provide a basis for risk assessment.

The component–functionality matrix is a way of mapping this relationship. Table 4.1 presents the relationship between components (presented in Step 3) and functionalities (presented in Step 1 and 2). According to Table 4.1, the measure function, for example, uses sensor, datalogging subsystems, and programming codes in its process. There is also a software functionality row that represents all software-related functionalities as presented in Step 1.

The analyst should expand this matrix according to the case under study and software application.

4.2.5 Step 5: Defining Software State Transitions Diagram

In this step, an explicit knowledge about the system behavior should be obtained, for example, by finite-state machines (FSM) method, [30, 31]. Modeling starts with defining the initial state and end state(s) of the software operation. Then, the modes of success and failure within the software should be identified. Each failure and success mode is given a name and will be considered as an end state. There are usually several modes through which the system progresses before reaching the end states. These modes will be considered as the transitional states. In this step, all possible paths from the initial state to end state(s) will be considered, and the functionalities that make the transitions possible are listed. The state transition diagram provides a basis for event sequence development. The development of these diagrams is enabled by the component–functionality matrix in the previous step.

Table 4.1 Component–Functionality matrix of the example software

	Convert	Open	Close	Regulate	Measure	Transfer data
Configuration system	X					
Unit converter	X					
Control system						
Shutdown controller		X	X			
Controller1		X	X			
Controller2		X	X			
Isolation controller		X	X	X		
Regulation controller		X	X	X		
Sensor					X	
Data logging system	X				X	X
Software functionality	X	X	X	X	X	X
Processor functionality	X	X	X	X	X	X
Value-related functionality	X	X	X	X	X	X
Time-related functionality	X	X	X	X	X	X
Interaction-related functionality	X	X	X	X	X	X

The visual representation of the transition of the system between states is the transition diagram. In Fig. 4.7, an example of the state transition between software states, and its environment (human and hardware systems' states) is presented. The state transition diagram for a software under study should be developed based on the software application and operating process. In Fig. 4.7, the first state is a software-related alarm. We could have a false alarm; thus, this event has an accuracy probability (we could also have a software failure without any alarm, that is not presented here).

The following states are developed according to the component–functionality matrix, and the last states are end states, which are safe (success) and fail, in this

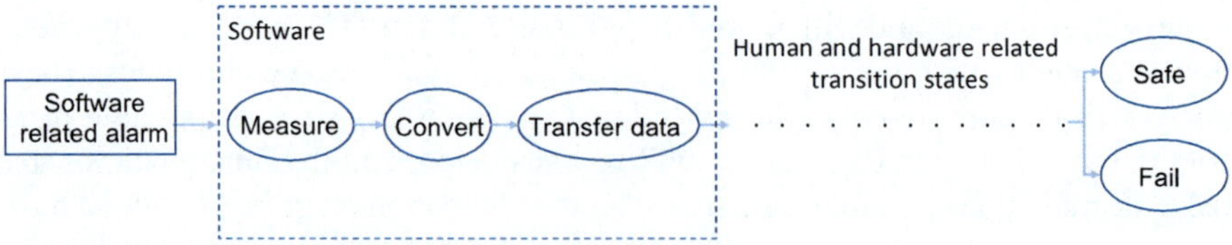

Fig. 4.7 State transition diagram after failure/error for the example software system

diagram. End states should be defined based on the application, system boundary, and goal of the study. For instance, here, safe is defined as the proper functionality of the software and involved components, presented in the state transition diagram (not overall system).

4.2.6 Step 6: Generating Software Event Sequence Diagram

In this step, the functionalities presented in the state transition diagram (Fig. 4.8) are replaced with their logical equivalents in terms of the subsystem/component states based on the level of the modeling. The goal is to generate Event sequence diagrams that represent accident scenarios, which could be further analyzed in risk assessments.

For example, when we have an alarm related to measurement malfunction, sensor, data logging subsystems, programming code subsystem/components, and software functionality (processor, value, time, and interaction-related functionalities) are involved (Fig. 4.2). Then, based on the system state and developed logic rules (Step 2), the probable faulty software functions are identified. In the presented example, we can assume that we have a faulty output value.[3] Thus, an event sequence diagram can be generated that includes sensor, data logging subsystems, and value-related failure events, as presented in Fig. 4.8.

The ESD generation follows the same rule for any other alarms. First, we should specify the type of alarm and malfunctions, then components involved to perform the specific function(s) will be defined (using component-functionality matrix). Finally, an ESD, presenting all subsystems/components, can be generated based on the state transition diagram. The analyst also decides how many times an event might be repeated in a scenario. This is necessary to prevent the ESD from getting trapped in infinite loops. In this step, it is assumed that events have binary states of success and failure. We can have more than two states for each event and have multiple end states accordingly, i.e., healthy, degraded, and failed.

The ESD can be part of probabilistic risk assessment, for example, linked to fault trees and BBNs as in HCL [32], or as a basis for simulation of scenarios (see Sect. 4.4).

4.2.7 Cybersecurity in Risk Assessment

Cyberattacks affect the risk level of autonomous systems. Cybersecurity is an important part of system safety, and safety is a requirement for security. Cybersecurity and risk control actions should be established and considered in the risk assessment of autonomous systems. In general, several methods for cyber risk and security are

[3] Based on the system state (C value) and the logic rule, presented in Fig. 1.5.

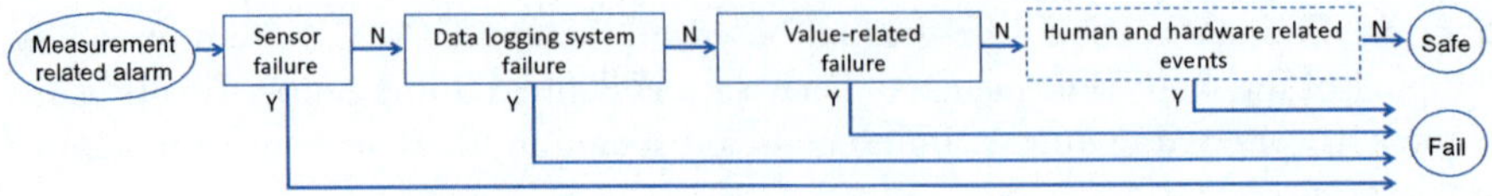

Fig. 4.8 Generated event sequence diagram for the example software (N: no, Y: yes)

available, for example, [33–35] which are related to different autonomous systems. Comprehensive risk management models should be developed, capable of optimizing design and operation of systems in order to decrease cyberattack risk level, considering system restrictions. A major challenge is to identify threats, potential attackers, and generate all possible scenarios related to cyberattacks.

Unknown unknowns are a challenge in all types of risk assessment, but for cybersecurity this is in particular relevant. There may be an infinite number of ways the system may fail since attackers can be so creative in finding ways to crash software. Hence, one way to deal with this challenge may be to limit the number of failure modes for software (or ways software may fail) and rather focus more on their impact, [36].

Autonomous ships, for example, are exposed to the threat of cyberattacks, due to their need to connect to operational, management, and administrative systems onshore [26]. The consequences of a cyberattack may not only impact the ship itself and/or its cargo, but also be a threat to the infrastructures along the coast. The most severe situation is if an unmanned and autonomous ship is hacked and control over it is lost partly or completely. In theory, boarding the ship may reestablish control (e.g., by helicopter), but this will take time in any case, and if the vessel is far from shore, a helicopter may not be able to reach the ship until it comes closer to shore, and then it may be too late to gain control [26]. A small crew onboard may be an effective risk control action.

4.3 Risk Assessment of Human and Organizational Factors

The operation of marine and maritime systems can be challenging, and the human operator may have to intervene. Such interventions require that the human operators have sufficient situation awareness to be able to take control and handle an emergency or difficulty. Research studies of dynamic positioning (DP) systems and collision risk for shuttle tanker—floating, production, storage, and offloading operations have shown that the DP operator may only have around 45 s to solve a problem before a collision occurs, which is too short time [37]. This gap between the expected improvement in operational performance through autonomous systems (e.g., the DP system) and the struggle of human operators to manage such advanced systems in the face of unique, unexpected events and emergencies makes it essential to analyze whether the technological drive toward higher autonomy improves safety and reliability of the marine operations under question. [38] uses high-reliability management

(HRM) to investigate the performance of DP and ROV control room operators, both facing complex systems and operations with different levels of autonomy, onboard an intervention vessel. This research is summarized below.

The foundation for HRM is based on a long-term study of the central control room at the California Independent System Operator (CAISO). CAISO is an organization responsible for managing the high-voltage electricity grid in California [39]. Later, control room research at other large-scale infrastructures, including water and marine systems, contributed to refining the framework (see, References [40] and [41]). The HRM framework describes the real-time management challenges when control room operators´ experience, knowledge, and management of systems must compensate for persisting limitations in design and technology.

A challenge with high autonomy is to enable cooperation and sufficient information flow between the human operator and the autonomous system within and across differing LoAs. Autonomy is a functionality that must allow the operator and the autonomous system to interact in a safe, reliable, and efficient manner [42]. Developing designs and technology that safely enable rather than limit shifts from one LoA to another, especially under sudden and/or unforeseen events faced by operators and system managers, is a major concern.

HRM is constituted by two main system dimensions, i.e., task volatility and options variety, as shown in Table 4.2. These dimensions are used to characterize the performance of control room operators during normal operations. The volatility refers to uncontrollable changes and unpredictable conditions that influence control room operators in complying with task demands (e.g., balancing load and generation on the electric transmission grid). Options' variety represents the available resources for meeting the task demand, such as ensuring balance between load and generation. The two dimensions use "high" and "low" for heuristic purposes, and the resulting four performance modes have different characteristics and safety envelopes for the control room operators.

Control room operators need to be able to maneuver between and across the performance modes to maintain normal operations and manage their changing risks. Key operational risks in each performance mode vary as option variety and system volatility change. Therefore, the performance modes can be used to identify risks related to a system.

The just-in-case performance mode occurs when operations are routine, and many management strategies and options are available, whereas the just-in-time implies improvisation of options and strategies when task conditions are more unstable. The just-for-now is when control operators may have to operate temporarily in a highly

Table 4.2 Performance modes of HRM in normal operations [39]

Dimensions		System volatility	
		High	Low
Option variety	High	Just-in-time performance	Just-in-case performance
	Low	Just-for-now performance	Just-this-way performance

challenging mode, i.e., when system instability is high and option variety are few. To stabilize the situation, just-this-way may occur in emergencies when options have dwindled and there is one "way" followed to solve the situation. Normal operations not going according plan are the usual course of things, but in addition, there are disrupted and failed operations.

From the HRM perspective, the disrupted and failed states of operations occur when the control operators no longer know in real time if options variety, nor task volatility, is "high" or "low" or anywhere in between. The difference between disrupted and failed operation is related to the time aspect, because the lack of real-time knowledge for control operators can be temporary (disruption) or indefinite (failure).

The performance modes of HRM have different dominant risks:

- Just-in-case performance: Complacency and inattentiveness to emerging or potential changes in system volatility or options variety.
- Just-in-time performance: Misjudgment by control operators with so many variables in the air at one time.
- Just-for-now performance: The most unstable performance mode and the one control operators and risk managers most want to avoid. The risk is losing maneuverability by tunneling into a course of actions without escape alternatives.
- Just-this-way performance: Some operator(s) fails to comply with command and control measures to reduce the challenging task volatility.

The risks of complacency, misjudgment, loss of maneuverability, and noncompliance are real-time risks and threaten system operations as a whole during normal operations. They include uncertainty that is difficult or impossible to measure in real time with fast-changing task volatility and options in the ocean environment. Hence, from the perspective of HRM, these four specific risks are imperative to mitigate in risk management of autonomous marine and maritime systems, both during the design process and in operations.

The demands on real-time risk management are acute for vessel control operators for several reasons. An intervention vessel, for example, has multiple automated systems (e.g., the remotely operated vehicle (ROV) and DP system of the vessel), each with various LoA. Further, normal operations are exposed to being disrupted and failed, which means that the risks associated with restoration or recovery of these systems can be severe, for example, if the DP system fails during a subsea intervention operation. A higher autonomy level means more time in the just-in-case performance mode for the human operator, but then the demand for the experienced operator is highest during unforeseen contingencies or crises when little time remains for to take effective action. Simulations and operator training are imperative if an operator is to be able to handle such conditions.

When designing systems with functionality in higher LoA, it is critical to assess the main risks of the performance modes, potential causes to these risks in the relevant system/operation, and further determine how the risks can be mitigated. In a hostile ocean environment, design transformations and management improvements have to focus on reducing control room operators´ internal task volatility limiting real-time interventions, which requires increasing the real-time decision options, including their decision support systems. Improved real-time decision support systems for human operators could improve their situation awareness and operational efficiency for safer autonomous functionalities.

Methods for quantification of human factors are presented in Chap.0 5.

4.4 An Online Risk Assessment Framework for Considering Human, Hardware, and Software Failure Interactions

An important challenge with risk assessment of complex systems is how to model cascading failures, and human, hardware, and software interactions.

A cascading failure is a process in a system of interconnected parts in which the failure of one or few parts can trigger the failure of other parts and so on. Cascading failures may occur when one part of the hardware or software of the system fails, or if a human error occur. When this happens, other parts including hardware, software, and human elements must compensate for the failed component. This in turn may overload these parts, causing them to fail as well, prompting additional parts to fail one after another [43, 44].

Sun et al. [45] has reviewed models of cascading failure in complex power grids. In this study, different models are presented to identify failure propagations in power grids. Other methods for different applications such as electric systems, urban infrastructure [46], critical infrastructures, and railways [47] have been proposed in last decades. For instance, in software risk assessment, the output and hence the effect or consequence of each failure mode on the external interfaces needs to be analyzed with respect to both effect on the "local" subsystem, but also on the overall system functionality and the context (globally). The critical aspect in this step is to identify and analyze how the software failure modes interact and impact the external interface through any propagation behavior. The importance of considering failure propagation is explained in [4].

Generally, the propagation of the effects resulting from the failure modes highly depends on the software function. The effect of control loops and reiterations within the software should be considered. The propagation should be reiterated at least once for feedback loops, such that the effect of these becomes visible. Faults in the feedback may not be apparent upon their occurrence, since the failure may occur after the feedback is used. Hence, the influence on the software functions that use the feedback needs to be assessed. Additional iterations may be necessary. Fault detection and correction mechanisms also need to be considered while analyzing the failure propagation behavior.

Table 4.3 presents an example of a propagation behavior for a specific type of failure mode. The complete propagation behavior of the failure modes through a software system is presented in [4].

Table 4.3 Propagation behaviors and effects for time-related failure modes for provided output, from [3, 4]

Failure mode	Propagation behavior and effect
Too early	No output is registered, propagates as no value failure mode
	Failure is detected, and the software aborts the operation
	Failure is masked; the premature value is stored in a buffer and available for further operation
Too late	Fault propagates as delayed output by the same time as the initial delay
	Delay is detected if it is longer than the programmed interval, software aborts operation
	If the delay is longer than the specified interval, a standard value/behavior is used that is propagated as incorrect value failure mode
Spuriously	No output is registered, propagates as no value failure mode. Alternatively, a spurious action is triggered that will propagate as too early failure mode

Figure 4.9 presents a general conceptual model for online risk assessment that is able to consider a system´s software, human, and hardware models' interactions, including impacts from its operating environment. The conceptual model is an enhancement of the Accident Dynamic Simulator (ADS) [48], which is an advanced tool for dynamic probabilistic risk assessment (DPRA) mainly developed and applied for nuclear power plants. The conceptual model presented here is developed for online risk assessment of autonomous systems in particular.

DPRA simulations may be considered an "exploration" process of hazardous scenarios, and hence it is believed that the DPRA can benefit from the incorporation of enhanced rules for the scheduling of sequence simulations (see Chapter 2). This can be performed using the "scheduler" and "planner" models, presented in Fig. 4.9.

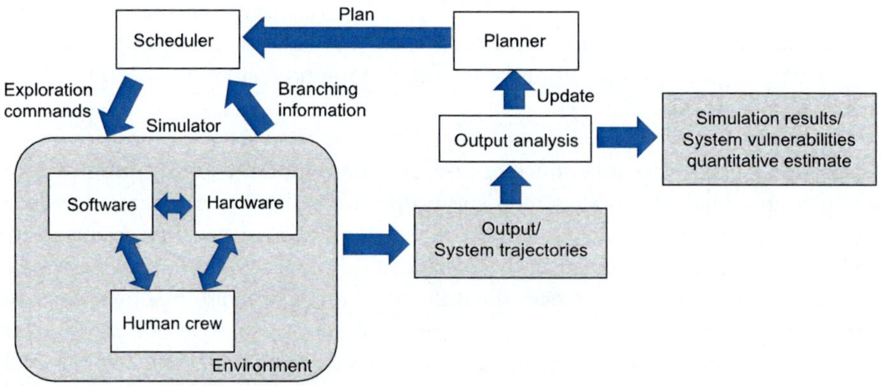

Fig. 4.9 A general conceptual model of hardware, human, and software interaction for online risk assessment of autonomous systems

A detailed description of this general conceptual model is presented in [49, 50]. The main components include:

The software–human–hardware is responsible for generating detailed system behavior as intended by system designers. The simulator consists of models to simulate the behavior of software, hardware, or human components of the system under study.

The scheduler manages the simulation process by saving system states, deciding the branch selection, adjusting the simulation levels of detail, and restarting the simulation. The scheduler guides the simulation toward the generated plan by the planner. The scheduler keeps track of the simulation and guides it adaptively.

The planner uses engineering knowledge to generate high-level scenarios. The planner guides the simulation by providing high-level scenarios to the scheduler. The generated scenarios could be incomplete or incorrect. The planner learns the detailed simulation results after a number of simulations runs and provides suggestions on possible ways to enhance the plan generated by the planner, for instance, by generating unseen scenarios or removing impossible ones [49, 51].

The "scheduler" is the main component of the conceptual method that represents software–human–hardware interdependencies. The scheduling is defined as the process of controlling the input to the simulation model, in order to simulate the desired types of scenarios. In the scheduler module, the branching rules are defined. The system simulator proposes the transitions to the scheduler whenever the simulator reaches a branching point. The scheduler then retrieves the value of the proposed transitions and decides which branch is to be explored. The exploration command is sent to the simulator for execution and the simulation continues until the next branching point or an end state is reached. A general concept of the scheduler is presented in several DPRA studies, such as for a nuclear power plant [52]. In these studies, a scheduler is used to consider the interactions among system hardware, software, and human behavior. In [53, 54], a detailed description of the implementation of schedulers in DPRAs is discussed and a nuclear power plant is presented as a case study.

Using the scheduler module for autonomous systems, in addition to the modeling the software–hardware–human interactions, environmental factors need to be evaluated. Weather conditions, for example, could affect human and hardware performance. For human performance, performance shaping factors (PSFs) are taken as input to the crew behavior model (see Chap. 5), which simulate operator(s) behavior in response to an initial event. For instance, in a DP system, training, procedures, rules, and restrictions impact the crew, and changes could affect crew behavior.

The operator(s) or crew actions affect system performance. On the other hand, any change in the system performance, influence operator(s)´ decision-making and action execution process.

Software failures could be identified by using the proposed method and taxonomies as described earlier; and potential failure propagation should be analyzed with respect to effects on hardware and human performance, providing important information to be used when generating the hazardous scenarios. In current DPRA, software failures and interactions are only explored to a limited extent.

Input parameters, such as environmental factors, are recorded in real-time operation in the simulator [55–61]. The crew and system simulation models are also run in real time. Furthermore, as presented, the interaction between the crew model and system simulation model is considered in real-time operation. As a result, the presented framework is capable of considering system performance and model interactions in real time, if the inputs are updated accordingly.

References

1. Skogdalen JE, Vinnem JE (2011) Quantitative risk analysis offshore—human and organizational factors. Reliab Eng Syst Saf 96(4):468–479
2. Parhizkar T, Utne IB, Vinnem JE, Mosleh A (2021) Dynamic probabilistic risk assessment of decision-making in emergencies for complex systems, case study: dynamic positioning drilling unit. Ocean Eng 237:109653
3. Thieme CA, Mosleh A, Utne IB, Hegde J (2020) Incorporating software failure in risk analysis–Part 1: Software functional failure mode classification. Reliab Eng Syst Safety 197:106803
4. Parhizkar T, Vinnem JE, Utne IB, Mosleh A (2020) Supervised dynamic probabilistic risk assessment of complex systems, part 1: general overview. Reliab Eng Syst Safety, p 107406
5. Parhizkar T, Utne IB, Vinnem JE, Mosleh A (2021) Supervised dynamic probabilistic risk assessment of complex systems, part 2: Application to risk-informed decision making, practice and results. Reliab Eng Syst Saf 208:107392
6. Tan S, Moinuddin K (2019) Systematic review of human and organizational risks for probabilistic risk analysis in high-rise buildings. Reliab Eng Syst Saf 188:233–250
7. Hogenboom S, Parhizkar T, Vinnem JE (2021) Temporal decision-making factors in risk analyses of dynamic positioning operations. Reliab Eng Syst Saf 207:107347
8. Skogdalen JE, Vinnem JE (2012) Combining precursor incidents investigations and QRA in oil and gas industry. Reliab Eng Syst Saf 101:48–58
9. Gran BA, Bye R, Nyheim OM, Okstad EH, Seljelid J, Sklet S, Vatn J, Vinnem JE (2012) Evaluation of the Risk OMT model for maintenance work on major offshore process equipment. J Loss Prevent Process Industries 25(3):582–593
10. Vinnem JE, Bye R, Gran BA, Kongsvik T, Nyheim OM, Okstad EH, Seljelid J, Vatn J (2012) Risk modelling of maintenance work on major process equipment on offshore petroleum installations. J Loss Prevent Process Industries 25(2):274–292
11. Bye A, Laumann K, Taylor C, Rasmussen M, Øie S, van de Merwe K, Øien K, Boring R, Paltrinieri N, Wærø I, Massaiu S (2017) The petro-HRA guideline
12. Dong Y, Rokseth B, Vinnem JE, Utne IB (2016) Analysis of dynamic positioning system accidents and incidents with emphasis on root causes and barrier failures. In: Proceedings of ESREL 2016 (Glasgow, Scotland, 25–29 September 2016), 166
13. Parhizkar T, Hogenboom S, Vinnem JE, Utne IB (2020) Data driven approach to risk management and decision support for dynamic positioning systems. Reliab Eng Syst Saf 201:106964
14. Arnhus M (2014) Modeling of technical, human and organizational factors and barriers in marine systems failure risk
15. Aven T, Hauge S, Sklet S, Vinnem JE (2006) Methodology for incorporating human and organizational factors in risk analysis for offshore installations. Int J Mater Struct Reliab 4(1):1–14
16. Thieme CA, Utne IB (2017) A risk model for autonomous marine systems and operation focusing on human–autonomy collaboration. Proc Inst Mech Eng Part O: J Risk Reliab 231(4):446–464
17. Grabowski M, You Z, Zhou Z, Song H, Steward M, Steward B (2009) Human and organizational error data challenges in complex, large-scale systems. Saf Sci 47(8):1185–1194

18. Wei Y, Rodriguez M, Smidts CS (2010) Probabilistic risk assessment framework for software propagation analysis of failures. Proc Inst Mech Eng Part O: J Risk Reliab 224(2):113–135
19. Holmberg JE (2014) Software reliability analysis in probabilistic risk analysis. In: Progress of nuclear safety for symbiosis and sustainability. Springer, Tokyo, pp 307–315
20. Smogeli O, Skogdalen JE (2011, May) Third party HIL testing of safety critical control system software on ships and rigs. In: Offshore technology conference. OnePetro
21. Ristord L, Esmenjaud C (2002) FMEA Performed on the SPINLINE3 operational system software as part of the TIHANGE 1 NIS Refurbishment Safety Case (No. NEA-CSNI-R--2002-1)
22. Huang B, Zhang H, Lu M (2009, July) Software FMEA approach based on failure modes database. In: 2009 8th international conference on reliability, maintainability and safety. IEEE, pp 749–753
23. Park GY, Kim DH, Lee DY (2014) Software FMEA analysis for safety-related application software. Ann Nucl Energy 70:96–102
24. Al-Dabbagh AW (2009) Dynamic flowgraph methodology for reliability modelling of networked control systems: with application to a nuclear-based hydrogen production plant, Doctoral dissertation, Master of Applied Science, Electrical and Computer Engineering, Ontario Tech University
25. Stamatelatos M, Dezfuli H, Apostolakis G, Everline C, Guarro S, Mathias D, Mosleh A, Paulos T, Riha D, Smith C, Vesely W (2011) Probabilistic risk assessment procedures guide for NASA managers and practitioners
26. Zhu D (2005) Integrating software behavior into dynamic probabilistic risk assessment (Doctoral dissertation)
27. Zhu D, Mosleh A, Smidts C (2007) A framework to integrate software behavior into dynamic probabilistic risk assessment. Reliab Eng Syst Saf 92(12):1733–1755
28. Diaconeasa MA, Mosleh A (2018) Performing an accident sequence precursor analysis with the ADS-IDAC dynamic PSA software platform
29. Tumer I, Smidts C (2010) Integrated design-stage failure analysis of software-driven hardware systems. IEEE Trans Comput 60(8):1072–1084
30. Delgoshaei P, Austin M (2012) Software patterns for traceability of requirements to finite state machine behavior. Proc Comput Sci 8:214–219
31. Lee E, Kim YG, Seo YD, Seol K, Baik DK (2018) RINGA: Design and verification of finite state machine for self-adaptive software at runtime. Inf Softw Technol 93:200–222
32. Røed W, Mosleh A, Vinnem JE, Aven T (2009) On the use of the hybrid causal logic method in offshore risk analysis. Reliab Eng Syst Saf 94(2):445–455
33. Guzman NHC, Kufoalor DKM, Kozine I, Lundteigen MA (2019, September) Combined safety and security risk analysis using the UFoI-E method: A case study of an autonomous surface vessel. In: Proceedings of the 29th European safety and reliability conference
34. Cui J, Sabaliauskaite G (2017) On the alignment of safety and security for autonomous vehicles. In: Proceedings of IARIA CYBER, pp 1–6
35. Koschuch M, Sebron W, Szalay Z, Török Á, Tschiürtz H, Wahl I (2019, November) Safety & security in the context of autonomous driving. In: 2019 IEEE international conference on connected vehicles and expo (ICCVE). IEEE, pp 1–7
36. Thieme C, Ramos MA, Utne IB, Mosleh A (2021) Proceedings of the international workshop on autonomous systems safety, online event
37. Vinnem JE, Liyanage JP (2008) Human-technical interface of collision risk under dynamic conditions: an exploratory learning case from the North Sea. Int J Technol Human Interact (IJTHI) 4(1):35–47
38. Utne IB, Schjølberg I, Roe E (2019) High reliability management and control operator risks in autonomous marine systems and operations. Ocean Eng 171:399–416
39. Roe E, Schulman PR (2008) High reliability management: operating on the edge, vol 19. Stanford University Press
40. Schulman P, Roe E (2016) Reliability and risk: the challenge of managing interconnected infrastructures. Stanford University Press

41. Roe E, Schulman PR (2018) A reliability & risk framework for the assessment and management of system risks in critical infrastructures with central control rooms. Saf Sci 110:80–88
42. Department of Defense (DoD) (2011) Unmanned Systems Integrated Roadmap FY2011-2036. Reference number: 11-S-3613
43. Parhizkar T, Aramoun F, Saboohi Y (2020) Efficient health monitoring of buildings using failure modes and effects analysis case study: air handling unit system. J Build Eng 29:101113
44. Parhizkar T, Aramoun F, Esbati S, Saboohi Y (2019) Efficient performance monitoring of building central heating system using Bayesian Network method. J Build Eng 26:100835
45. Sun K, Han ZX, Cao YJ (2005) Review on models of cascading failure in complex power grid. In: Power system technology, 13, 1–9. Chicago
46. Little RG (2010) Managing the risk of cascading failure in complex urban infrastructures. Disrupted cities: When infrastructure fails, 27–39
47. Adams K, Heidarzadeh M (2021) A multi-hazard risk model with cascading failure pathways for the Dawlish (UK) railway using historical and contemporary data. Int J Disaster Risk Reduct 56:102082
48. Hsueh KS, Mosleh A (1996) The development and application of the accident dynamic simulator for dynamic probabilistic risk assessment of nuclear power plants. Reliab Eng Syst Saf 52(3):297–314
49. Hu Y, Parhizkar T, Mosleh A (2021) Guided Simulation for Dynamic Probabilistic Risk Assessment of Complex Systems: Concept, Method, and Application Reliab Eng Syst Saf, p 108047
50. Maidana RG, Parhizkar T, Thieme ChA, Ramos M, Utne I, Mosleh A (2021) Towards risk-based autonomous decision-making with accident dynamic simulation, ESREL Conference, France
51. Nejad HS, Parhizkar T, Mosleh A (2021) Simulation based probabilistic risk assessment (SIMPRA): risk based design. 31st European Safety and Reliability Conference, 19–23 September 2021, Angers, France
52. Coyne K, MOSLEH A (2014) Nuclear plant control room operator modeling within the ADS-IDAC, version 2, dynamic PRA environment: part 1-general description and cognitive foundations. Int J Perform Eng 10(7)
53. Diaconeasa MA (2017) Integration of qualitative and quantitative hybrid causal logic into a simulation-based platform for probabilistic risk assessment of nuclear power plants. PhD thesis, University of California, Los Angeles
54. Parhizkar T, Mosleh A (2022) Guided probabilistic simulation of complex systems toward rare and extreme events. The 68th Annual Reliability & Maintainability Symposium (RAMS), Tucson, Arizona
55. Parhizkar T, Roshandel R (2017) Long term performance degradation analysis and optimization of anode supported solid oxide fuel cell stacks. Energy Convers Manage 133:20–30
56. Roshandel R, Parhizkar T (2016) Degradation based optimization framework for long term applications of energy systems, case study: solid oxide fuel cell stacks. Energy 107:172–181
57. Parhizkar T, Hafeznezami S (2018) Degradation based operational optimization model to improve the productivity of energy systems, case study: Solid oxide fuel cell stacks. Energy Convers Manage 158:81–91
58. Roshandel R, Parhizgar T (2013) A new approach to optimize the operating conditions of a polymer electrolyte membrane fuel cell based on degradation mechanisms. Energy Syst 4(3):219–237
59. Parhizkar T (2018) Long-term degradation-based modeling and optimization framework. In Handbook of research on predictive modeling and optimization methods in science and engineering (pp 192–220). IGI Global
60. Parhizkar T, Mosleh A, Roshandel R (2017) Aging based optimal scheduling framework for power plants using equivalent operating hour approach. Appl energy 205:1345–1363
61. Sotoodeh AF, Parhizkar T, Mehrgoo M, Ghazi M, Amidpour M (2019) Aging based design and operation optimization of organic rankine cycle systems. Energy Convers Manage 199:111892

Chapter 5
Human Factor Analysis and Quantification

5.1 Background

Cognitive reliability error analysis method (CREAM), a technique for human event analysis (ATHEANA), and standardized plant analysis risk-human reliability analysis (SPAR-H) are examples of static human reliability assessment (HRA) methods. In CREAM, performance conditions originating from common modes must define cognitive and contextual control modes, which, in turn, will increase or decrease the basic probabilities of each basic type of human error [1]. ATHEANA is oriented toward reliability analysis and could be included in probabilistic safety/risk assessment (PSA, PRA) event trees [2]. The SPAR–H method decomposes probability into contributions from diagnosis and action failures, accounts for the context associated with human failure events by using performance shaping factors (PSFs), and dependency assignment to adjust a base-case human error probability (HEP), guidance on how to assign the appropriate value of the PSF and offers an adjustment factor to reduce double counting of influencing factors shared by PSFs.

As human behavior is time-dependent, dynamic methods could reflect human errors more effectively. Dynamic logical analytical methodology (DYLAM), cognitive environment simulation (CES), integrated reactor operator/system (INTEROPS), cognitive simulation model (COSIMO), cognitive and action modeling of an erring operator (CAMEO), simulation system for behavior of an operating group (SYBORG) and simulation-based evaluation and analysis support system for man–machine interface design (SEAMAID) are examples of dynamic methods to evaluate human–machine interface (HMI) performance, and for human error identification [3, 4].

In [5], the information, decision, and action (IDA) cognitive model was developed. In [6], an information decision and action crew (IDAC) cognitive model was developed for nuclear power plants. This model is coupled to the accident dynamics simulator (ADS) [7–9], and dynamic event trees could be generated using this method. The following subsections outline these methods more in detail.

T. Parhizkar et al., *Online Probabilistic Risk Assessment of Complex Marine Systems*, Springer Series in Reliability Engineering,
https://doi.org/10.1007/978-3-030-88098-9_5

5.2 Understanding Operators/Crew Tasks

An important step when working with human factors in complex systems is to understand the organizational chart of the crew positions involved. In an organizational chart, we should have three main elements.

1. Crew member classification
2. Job description and duties of crew positions
3. The interactions between crew members.

For instance, according to the reviewed incident reports [10–13], in most of the dynamic positioning (DP) systems on offshore drilling vessels, there are eight classifications of operator(s) positions, presented in Table 5.1.

The duties of each operator should be derived from manuals/standards [14], interviews and/or questionaries provided by operator(s). For instance, offshore installation managers (OIM) have the following tasks and duties [15]:

- Ensuring full compliance with all mandatory regulations and procedures relating to production operations
- Overseeing the daily production/platform operation activities
- Managing emergencies
- Organizing, supervising and delegating personnel activities to ensure that preventative maintenance, housekeeping, and repairs are performed promptly and efficiently
- Promoting and maintaining a safe environment
- Making sure that the ongoing development and training of employees is up to date, and equipment is maintained to a high standard.

The dynamic positioning operator (DPO) reports to the senior DPO and is responsible for navigation and controlling the vessel's position, using the DP systems, manual controls, and joysticks as appropriate. The DPO assists the senior DPO in taking immediate action in any circumstance that could compromise the safety of the vessel, or the DP operation. The DPOs must be familiar with the latest failure modes

Table 5.1 Operator(s) classification for dynamic positioning (DP) system operation of a drilling vessel

No	Classification
1	Offshore installation manager
2	Marine section leader and dynamic position operator (DPO)
3	Admin and rig operations superintendent
4	Storekeeper
5	Technical section leader
6	Electrical-Instrument-Tele section lead
7	Drilling section leader
8	Process section leader

and effect analysis (FMEA) and annual DP trials, and the DP capability analyses and verification footprint. They must also be fully conversant with the configuration of the vessel's power management system.

The DPOs need to maintain a comprehensive understanding of the DP control systems, including procedures for both automatic and manual system changeovers, and have a comprehensive knowledge and detailed understanding of the vessel's DP operations manual, the company's field DP manual, and master/OIM's standing instructions. The DPO should ensure that all checklists are maintained, as detailed in the company's DP operations manuals. The DPO is also responsible for ensuring detailed familiarity with regulations, industry guidance and company procedures in respect of their duties. They also assist in the coordination of all deck, crane, gangway, helicopter and supply boat operations [16].

The tasks and duties of the rest of operator(s), presented in Table 5.1, should be derived. Then, the interaction between operator(s) could be presented in an organizational chart. Figure 5.1 illustrates an example of a possible organizational chart for a DP drilling vessel. There could be other organizational charts for DP drilling vessels. The letter presented in parenthesis in each box indicate the main type of task of the corresponding operator. An OIM mainly performs decision-making. The lines show the interaction between operator(s). For instance, as presented in the figure, the OIM communicates with admin, rig operations superintendent, and onshore personnel.

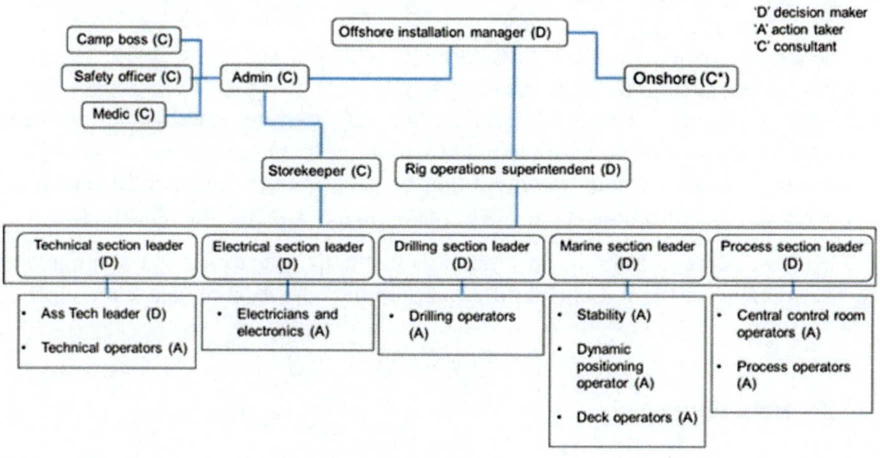

Fig. 5.1 Organizational chart of a drilling DP vessel

5.3 Performance Shaping Factors' Identification

Performance shaping factors (PSFs) are factors that may enhance or degrade human performance [17, 18]. These factors provide a basis for considering potential influences on human performance, and they need to be systematically considered in the quantification of human error probabilities (HEPs) [19, 20], to be presented later in this chapter. PSFs are often categorized as internal and external factors. The internal factors are the influences that the individual brings to the situation, such as mood, fitness, stress level, etc. The external factors are the situational or environmental influences that affect the individual, such as temperature, noise, and work practices. In the following, some of the important PSFs are presented, namely training and experience, procedures, ergonomics and human–machine interface (HMI), available time, task complexity, stress level, fitness for duty, and work processes.

5.3.1 Training and Experience

This PSF refers to the level of experience and training of the operator(s) involved in a task. The main factors considered in this PSF factor are:

- Years of experience of the crew or individual
- Training level of the specific task
- The amounts of time passed since training
- Systems involved in the task and scenario
- Novel or unique scenarios (i.e., whether or not the crew or individual has been involved in a similar scenario, in either a training or an operational setting) [21]. Specific examples where training might be deficient are guidance for manually operating/disconnection of the vessel (using joystick or manual disconnection).

There are different trainings in simulators that help operators to get familiar with the environment of the vessel and different possible operation modes and incidents.

5.3.2 Procedures

This PSF refers the existence and use of formal operating procedures for the tasks under consideration. Common problems with formal procedures include:

- Situations where procedures provide inadequate or wrong information related to a particular control sequence.
- Ambiguity of task steps.

Evaluation of this PSF should consider whether the activity is an action or a diagnosis activity [21].

5.3.3 Ergonomics and Human–Machine Interface (HMI)

Ergonomics is mainly about interaction and tasks of the operator/crew with the equipment [22], such as the control, display and equipment, and quality and quantity of information available from instrumentation.

Accident reports of collisions between attendant vessels and offshore oil and gas platforms on the Norwegian continental shelf have shown that, when necessary, indicators of system status are not located in one designated place on the HMI, and it is difficult for an operator to monitor all necessary information to properly control the facility [22]. This is one example of poor ergonomic related to panel design layout. If there is evidence that this is the case, a negative PSF value should be assigned.

Examples of poor labeling include instances where labels are temporary, informal, or illegible. In addition, multiple names may be given to the same piece of equipment [21].

5.3.4 Available Time

Available time reflects the amount of time that a crew or an operator has to diagnose and act based on an abnormal event. A lack/shortage of time can result in failure of operator to think clearly, consider alternatives, and it can affect her/his performance [21].

5.3.5 Task Complexity

Complexity refers to difficulty level of a task that should be performed in the given context. In addition to the task, complexity considers the environment in which the task is to be performed.

5.3.6 Stress Level

Stress refers to the level of undesirable conditions and circumstances that delay/impede the operator from completing a specific task. Stress can include excessive workload, mental stress, or physical stress. High stress level could result in narrowed attention, muscular tension, apprehension or nervousness associated with the event.

Environmental factors such as poor ventilation, excessive heat, noise, or radiation, can increase the stress in a person and affect his/her mental or physical performance.

It should be noted that the effect of stress on human performance is curvilinear—some small amount of stress can improve performance and is considered nominal, while high levels of stress are negatively affected human performance [21].

5.3.7 Fitness for Duty

Fitness for duty reflects how the individual is fit physically and mentally to perform the task at the time. Factors that affect this PSF include personal problems, fatigue, sickness, overconfidence, drug use, and distractions [21].

5.3.8 Work Process

Work processes refer to aspects of performing the work, such as team dynamics, interorganizational impacts, available staffing, work planning, safety culture, communication, and management policies.

5.3.9 Evaluating PSFs

Performance shaping factors could have different levels, and they could be quantified according to the definitions provided in standards. For instance, available time could have four levels as follows [21].

- Inadequate time: Operator(s) does not have enough time to perform the task, and it has extremely high negative effect on performance.
- Barely adequate time: The operator(s) has limited time to perform the task. However, there are more time than the minimum time required. In this situation, the operator(s) experiences high time pressure or that they have to speed up much to do the task in time. It has moderate negative effect on performance.
- Nominal time: There is enough time to do the task. The operator(s) only experiences a low degree of time pressure or need to speed up much to do the task. When comparing the available time to the required time, the analyst concludes that time would neither have a negative nor a positive effect on performance.
- Extra time: There is good time/extra time to perform the task. In this situation, the operator(s) has considerable extra time to perform the task, and there is no time pressure or need to speed up to do the task in time. It has moderate positive effect on performance.

The level of all PSFs should be defined for crew members of a complex system. It should be noted that PSF levels could change over time, and they should be updated

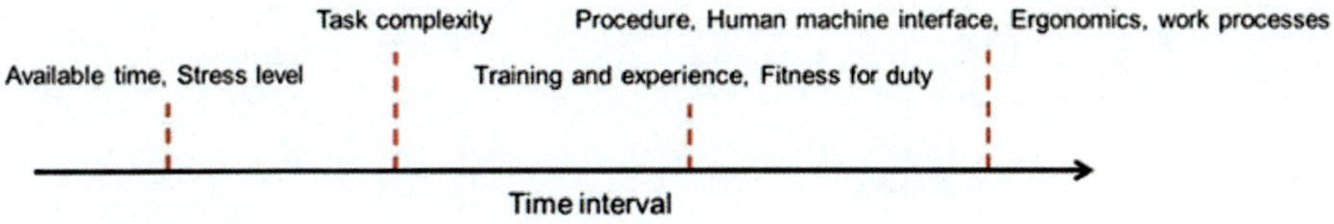

Fig. 5.2 Timelines of the updating process of PSFs

accordingly when used in online risk assessment. The timeline of updating PSF factors is different, and it depends on the nature of the factor.

Figure 5.2 presents relative timelines for different PSF factors. For instance, available time changes in real-time; thus, it has a very short time interval. However, fitness for duty, training and experience are updated whenever operators' shifts change; therefore, they have longer time intervals. There are other PSFs, mostly organizational related factors that have a longer updating timeline, such as procedures and human–machine interface.

5.4 Skill-Based Human Reliability Analysis—A DP Example

One of the powerful methods in human error quantification is skill-based human reliability analysis using Bayesian networks (BNs). In this method, human errors are categorized as slip, rule-based, and knowledge-based mistakes [23].

Slip errors are forgetfulness or involuntary mistakes. Rule-based mistakes are due to the choice of the wrong rule as a result of an erroneous perception of the situation or omissions in the application of a rule. These mistakes could happen when operators are guided by rules and procedures. The knowledge-based mistakes are due to lack of knowledge or incorrect application. These mistakes happen when the operators are in the presence of new or unforeseen situations, that is not known, for which they do not know the rules or procedures of reference. These errors could happen when operators must react creatively and independently, i.e., without the use of procedures or instinctive behaviors, based on the information available and the knowledge gained.

Each of these three categories is influenced by human skills and PSFs, which should be considered to determine the structure of the BN. The conditional probabilities and required inputs to the BN could be collected from historical data and expert judgments.

As an example, Fig. 5.3 shows a BN of operator(s) in a DP system of a drilling unit. At the first step, available incident reports from international marine contractor's association (IMCA) (see Chap. 3, Sect. 3.2) of DP systems are studied [24]. The analysis of the incidents from 2004 to 2015 indicates that human errors could be classified as slip, rule-based and knowledge-based mistakes. The PSFs are operators'

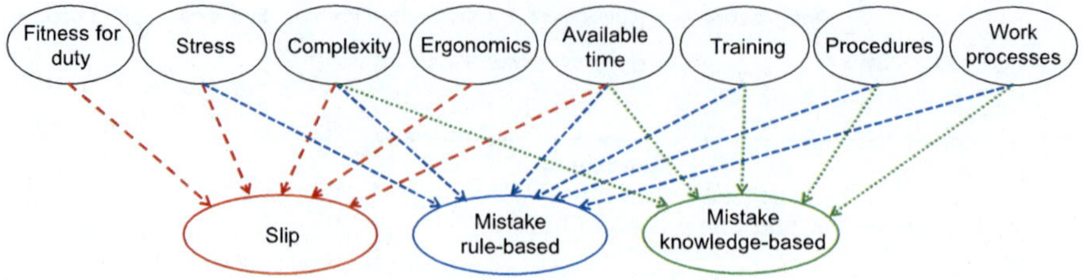

Fig. 5.3 Bayesian network of human error for a DP drilling unit

fitness for duty, stress, complexity, ergonomics, available time, training, procedures and work processes.

The child nodes of the BN could take discrete or continuous values. In the presented example, it is assumed that the parent nodes take binary values as presented in Table 5.2. In Table 5.2, PSFs of each state are presented as a multiplier. Multiplier values are derived from [21]. Since the PSFs are factors that could enhance or degrade human performance, they provide a basis for evaluating potential influences on human performance and for quantification of human error probabilities (HEPs), [20, 25].

The conditional probabilities between the BN nodes are calculated using the SPAR-H method as presented in [21]. In this method, the conditional probability of each state of nodes is equal to:

Table 5.2 Performance shaping factors (PSFs) of the dynamic positioning system for a drilling vessel [21]

PSFs	PSF levels	Multiplier
Available time	Inadequate time	P(failure) = 1
	Nominal	1
Stress	High	2
	Nominal	1
Complexity	Highly complex	5
	Nominal	1
Training	High	0.5
	Nominal	1
Procedures	Available, but poor	5
	Nominal	1
Ergonomics	Poor	10
	Nominal	1
Fitness for duty	Unfit	P(failure) = 1
	Nominal	1
Work processes	Poor	2
	Nominal	1

Table 5.3 Conditional probability table of the human error "slip"

Fitness for duty	Available time	Complexity	Ergonomics	Stress	Probability
Fit	Nominal	Nominal	Nominal	Nominal	0.00
				High	0.02
			Poor	Nominal	0.10
				High	0.19
		High	Nominal	Nominal	0.05
				High	0.10
			Poor	Nominal	0.38
				High	0.59

$$CP = \frac{0.01 \times \prod PSFs}{0.01 \times \left(\prod PSFs - 1\right) + 1} + \frac{0.001 \times \prod PSFs}{0.001 \times \left(\prod PSFs - 1\right) + 1} \tag{5.1}$$

where PSF is the multiplier, presented in Table 5.2. The first term presents the conditional probabilities of a detection and diagnosis tasks failure, and the second term is the conditional probabilities of action task failure. The conditional probabilities of the slip, rule-based mistake, and knowledge-based mistake nodes of the BN are calculated using Eq. (5.1), and presented in Tables 5.3, 5.4 and 5.5, respectively. It should be noted that for inadequate available time and unfit fitness for duty, the final conditional probabilities are assigned 1, regardless of the other PSFs, i.e., lack of available time and fitness for duty results in failure probabilities equal to one (Table 5.2).

The presented human errors, i.e., slip, rule-based and knowledge-based mistakes, could affect the performance of the DP systems. The impact of the human errors varies for different systems. The systems' failure probabilities due to human error are quantified using the presented BN in Fig. 5.4. These failure probabilities could be fed into systems' fault trees, presented in Chap. 2 (Sect. 2.3.3), to consider human error probabilities in each system.

The conditional probabilities of human errors resulting in DP reference, control, propulsion, and power systems failure are calculated based on gathered information from IMCA reports and are presented in Tables 5.6, 5.7, 5.8 and 5.9, respectively. The conditional probabilities are used in the BN to evaluate failure of reference, control, thruster and power system probabilities due to human error. The derived values are used in the human error basic events in the fault trees of these systems, e.g., in Figs. 5.2, 5.3 and 5.4, Chap. 2, the human error is presented as a basic even of the power system fault tree. The quantity of this basic event is equal to the node probability of thruster system, derived from the presented Bayesian network (Fig. 5.4), [26].

At this point, the general BN structure and the related conditional probabilities are derived. During system operation, available data can be used to set evidence, and then—based on the conditional probabilities—the probabilities of the end states

Table 5.4 Conditional probability table of the human error "rule-based mistake"

Available time	Complexity	Procedure	Stress	Training	Work processes	Probability
Nominal	Nominal	Nominal	Nominal	Nominal	Nominal	0.00
					Poor	0.02
				High	Nominal	0.01
					Poor	0.01
			High	Nominal	Nominal	0.02
					Poor	0.04
				High	Nominal	0.01
					Poor	0.02
		Available, but poor	Nominal	Nominal	Nominal	0.05
					Poor	0.10
				High	Nominal	0.03
					Poor	0.05
			High	Nominal	Nominal	0.10
					Poor	0.19
				High	Nominal	0.05
					Poor	0.10
	High	Nominal	Nominal	Nominal	Nominal	0.05
					Poor	0.10
				High	Nominal	0.03
					Poor	0.05
			High	Nominal	Nominal	0.10
					Poor	0.19
				High	Nominal	0.05
					Poor	0.10
		Available, but poor	Nominal	Nominal	Nominal	0.23
					Poor	0.38
				High	Nominal	0.12
					Poor	0.23
			High	Nominal	Nominal	0.38
					Poor	0.59
				High	Nominal	0.23
					Poor	0.38

Table 5.5 Conditional probability table of the human error "knowledge-based mistake"

Available time	Complexity	Procedure	Training	Work processes	Probability
Nominal	Nominal	Nominal	Nominal	Nominal	0.00
				Poor	0.02
			High	Nominal	0.01
				Poor	0.01
		Available, but poor	Nominal	Nominal	0.05
				Poor	0.10
			High	Nominal	0.03
				Poor	0.05
	High	Nominal	Nominal	Nominal	0.05
				Poor	0.10
			High	Nominal	0.03
				Poor	0.05
		Available, but poor	Nominal	Nominal	0.23
				Poor	0.38
			High	Nominal	0.12
				Poor	0.23

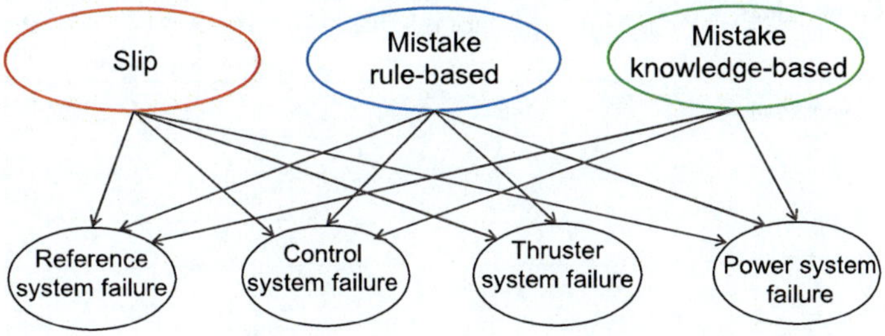

Fig. 5.4 Bayesian network for quantification of systems' failure probabilities due to human error

Table 5.6 Conditional probability of human error, resulted in dynamic positioning reference system failure

Slip	Mistake rule-based	Mistake knowledge-based	Probability
Yes	Yes	Yes	0.47
		No	0.36
	No	Yes	0.30
		No	0.15
No	Yes	Yes	0.37
		No	0.24
	No	Yes	0.17
		No	0

Table 5.7 Conditional probability of human error, resulted in dynamic positioning control system failure

Slip	Mistake rule-based	Mistake knowledge-based	Probability
Yes	Yes	Yes	0.81
		No	0.63
	No	Yes	0.75
		No	0.53
No	Yes	Yes	0.59
		No	0.22
	No	Yes	0.48
		No	0

Table 5.8 Conditional probability of human error, resulted in propulsion system failure

Slip	Mistake rule-based	Probability
Yes	Yes	0.24
	No	0.09
No	Yes	0.16
	No	0

Table 5.9 Conditional probability of human error, resulted in dynamic positioning power system failure

Slip	Mistake rule-based	Mistake knowledge-based	Probability
Yes	Yes	Yes	0.66
		No	0.49
	No	Yes	0.48
		No	0.21
No	Yes	Yes	0.58
		No	0.35
	No	Yes	0.35
		No	0

will be derived. Whenever there is an update to the input data, the process should be repeated, and new failure probabilities will be generated. This process is presented in Fig. 5.5.

5.5 Cognitive-Based Human Reliability Analysis

So far in the chapter, the focus has been on evaluating human errors and quantifying their probabilities. A comprehensive risk assessment with human factors requires a cognitive engine to model the operator decision-making and action execution

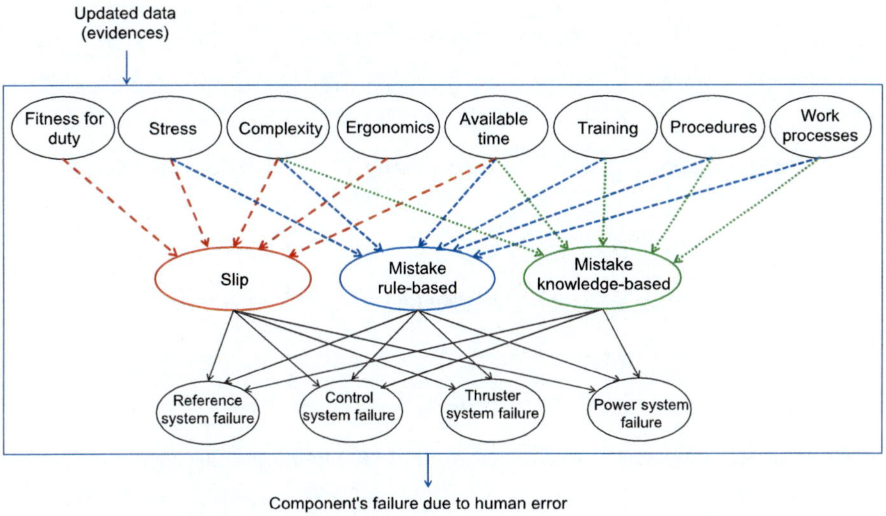

Fig. 5.5 Bayesian network for quantification of components' failure probabilities due to human error based on available evidence from IMCA reports [24]

processes. A cognitive engine is based largely on a recognition primed naturalistic decision-making (RPD) model [27–29]. Multiple cognitive engines are developed in the last decades [30–33].

The information, decision, and action in a crew context cognitive model (IDAC) is one of the powerful cognitive engines that is connected to the system accident dynamic simulator (ADS) [9]. This engine is developed for dynamic probabilistic risk assessment of nuclear power plants. In this section, the IDAC method is generalized to be applicable to any complex systems such as DP. The IDAC model considers the interaction between human operators' behavior and system technical factors, and it consists of three main modules, namely information filtering, decision-making, and action execution, as presented in Fig. 5.6.

As can be seen in Fig. 5.6, the IDAC cognitive model includes the information filtering, decision-making, and action execution modules. These modules model operators behavior in response to an incident in a complex system. PSFs enter the mental state module and based on the status of information filtering, decision-making and action execution modules; the mental state will be derived. The mental state and system technical and environmental factors enter information filtering module. Information filtering, decision-making, and action execution could be modeled with different levels of detail, depending on the available information and the goal of the study. In this chapter, IDAC and Bayesian network are presented as examples of a highly detailed method and a more simplified approach, respectively.

The highly detailed IDAC method can be used to have a better understanding of the human mental state over time, as the system is operating. In this section, the flow of data in information filtering, decision-making and action execution is

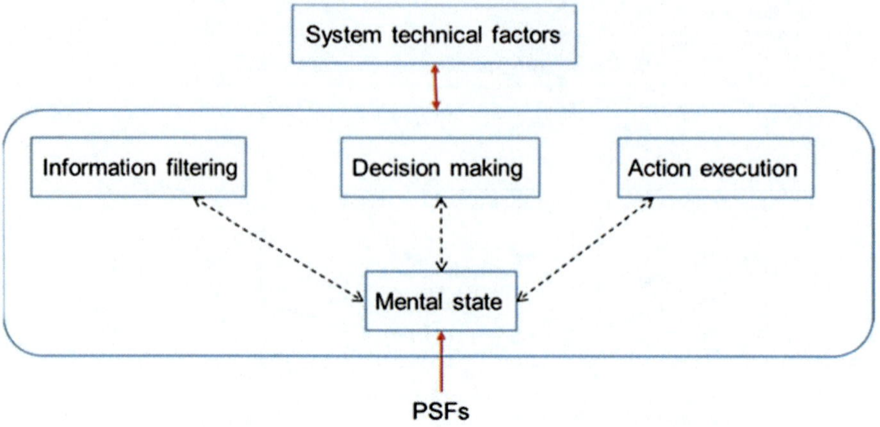

Fig. 5.6 Main components of IDAC cognitive model (PSF: performance shaping factor)

presented through multiple flowcharts. The quantification process of these flowcharts is presented in [34]. The process of information filtering is presented in Sect. 5.1, and the generalized IDAC flowcart is presented in Fig. 5.7, and related BNs are presented in Figs. 5.8 and 5.9. The output of information filtering, mental state and system technical and environmental factors enter decision-making module. The process of decision-making is presented in Sect. 5.2, and the IDAC flowchart and the BN are presented in Figs. 5.10 and 5.11. The output of the decision-making model, mental state, system technical and environmental factors enters the action execution module. The process of action execution is presented in Sect. 5.3.

It should be noted that the presented flowcharts in this Chapter specify the possible human behavior after an initial evet. Using these flowcharts, possible human behavior scenarios could be generated. These scenarios are used to develop human behavior event sequence diagrams (ESDs) and connect the human behavior ESDs to the system and software event sequence diagrams. Integrating human behavior ESDs with system and software ESDs are briefly explained in Chap. 4 (Sect. 4.4).

5.5.1 Information Filtering

Information filtering is the first step of human behavior modeling. Operator(s) gather information, then sort the information based on its origin and the type of information. Afterward, the initial information would be processed. Figure 5.7 shows the generalized information processing flow diagram based on the IDAC model.

Operator(s) retrieve data from the communication board (block 2, Fig. 5.7). The processing is paused until new information is received that require making decisions or taking actions (blocks 3 and 4, Fig. 5.7). Then, based on source of data, the analysis of the data would be performed by operators (block 5, Fig. 5.7).

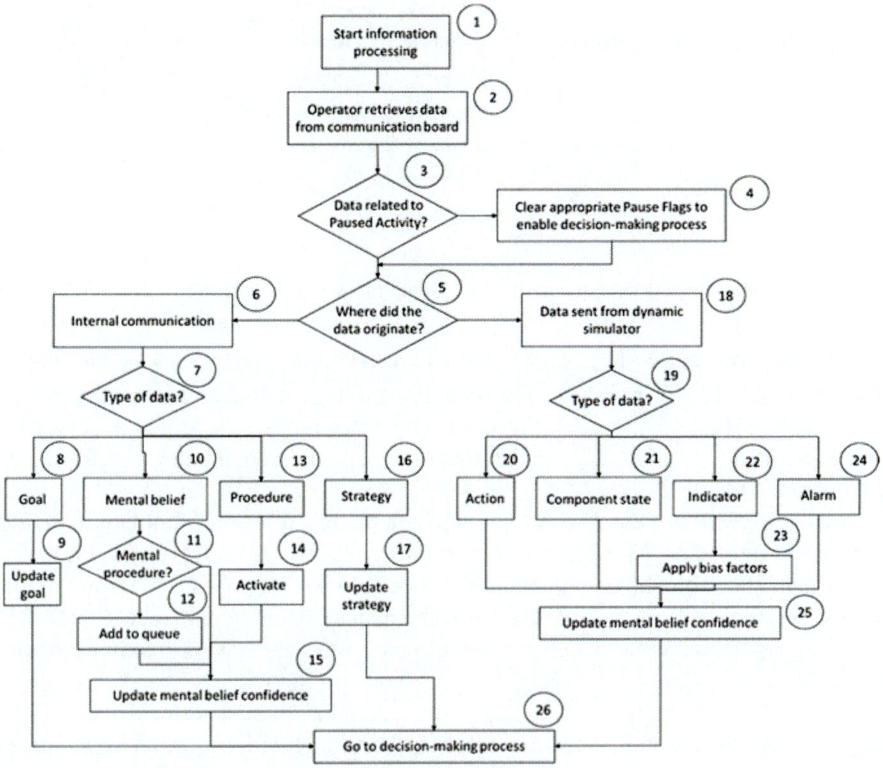

Fig. 5.7 Information processing flowchart of human operators

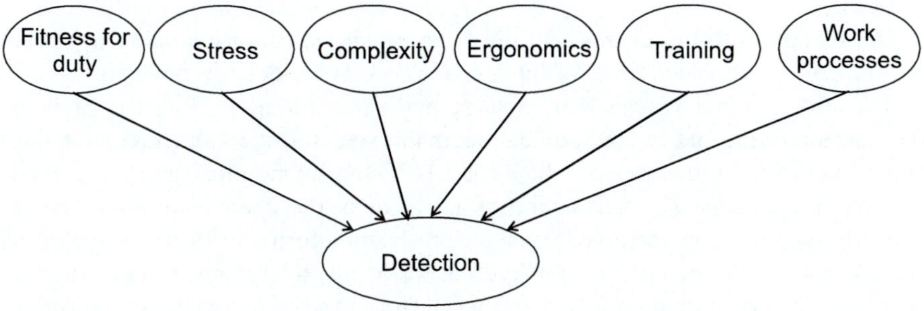

Fig. 5.8 Bayesian network of detection failure probability of a DP operator

Data retrieved from internal communication (block 6) could be categorized as goal (block 8, Fig. 5.7), mental belief (block 10, Fig. 5.7), procedure (block 13, Fig. 5.7), and strategy (block 16, Fig. 5.7). If data is related to the goal of the operation, then data could affect the operator goal in performing decision or taking an action (block 9, Fig. 5.7). The data related to the mental belief, i.e., the true or false belief

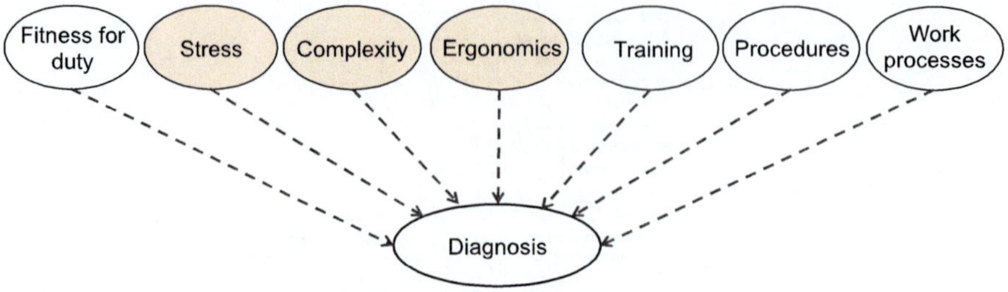

Fig. 5.9 Bayesian network of diagnosis failure probability of a DP operator

that the control system, has caused the loss of position, updates the mental belief (blocks 10, 11, 12, 15, Fig. 5.7). The operation strategy data updates the strategy of the operator (blocks 16, 17, 26, Fig. 5.7). The procedure data could activate a new procedure (block 14, Fig. 5.7), and all these data update mental belief confidence of the operator (block 15, Fig. 5.7).

Data retrieved from a simulator, which performs dynamic simulation of the complex system, such as ADS, could have four types of actions (block 20, Fig. 5.7), component state (block 21, Fig. 5.7), indicator (blocks 22, 23, Fig. 5.7), and alarm (block 24, Fig. 5.7). All these data update mental belief confidence (block 25, Fig. 5.7). Updated mental beliefs are used as inputs to the decision-making process, (block 26, Fig. 5.7), [34].

In [35], the information processing is considered as two stages of detection and diagnosis. At the first stage, the operator should detect deviations from normal operation. The failure probability of detection depends on different factors, such as fitness for duty, stress and complexity and could be calculated based on the BN presented in [35]. Figure 5.8 shows an example of a BN for detection failure probability in DP systems.

One of the factors that changes during an emergency situation compared to normal operation is the allocated time for the human operators to detect a deviation.

The next step is diagnosis that could be performed based on different methods, such as monitoring and/or communication. In this step, the operator tries to diagnose faults and errors in the system, which for a DPO means monitoring vessel screens and the environment. After detection of an incident, the operator monitors deviations based on the data perceived from one or several information sources, including alarm/event list, vessel position plot, hydroacoustic position reference system screen, position reference system information from console, thruster output, power consumption, vibration of bridge and acoustic noise (depending on vessel size and bridge location). In addition, the operator(s) will actively search and process the information in order to perform fault diagnosis.[1] According to these data, the operator(s) could determine likely failures and the severity of the situation. The probability of diagnosis failure depends on multiple factors and could be calculated using the Bayesian Network (BN) model, as presented in Fig. 5.9 for a DP system. The highlighted

[1] The information filtering step in IDAC method covers the detection and diagnosis process.

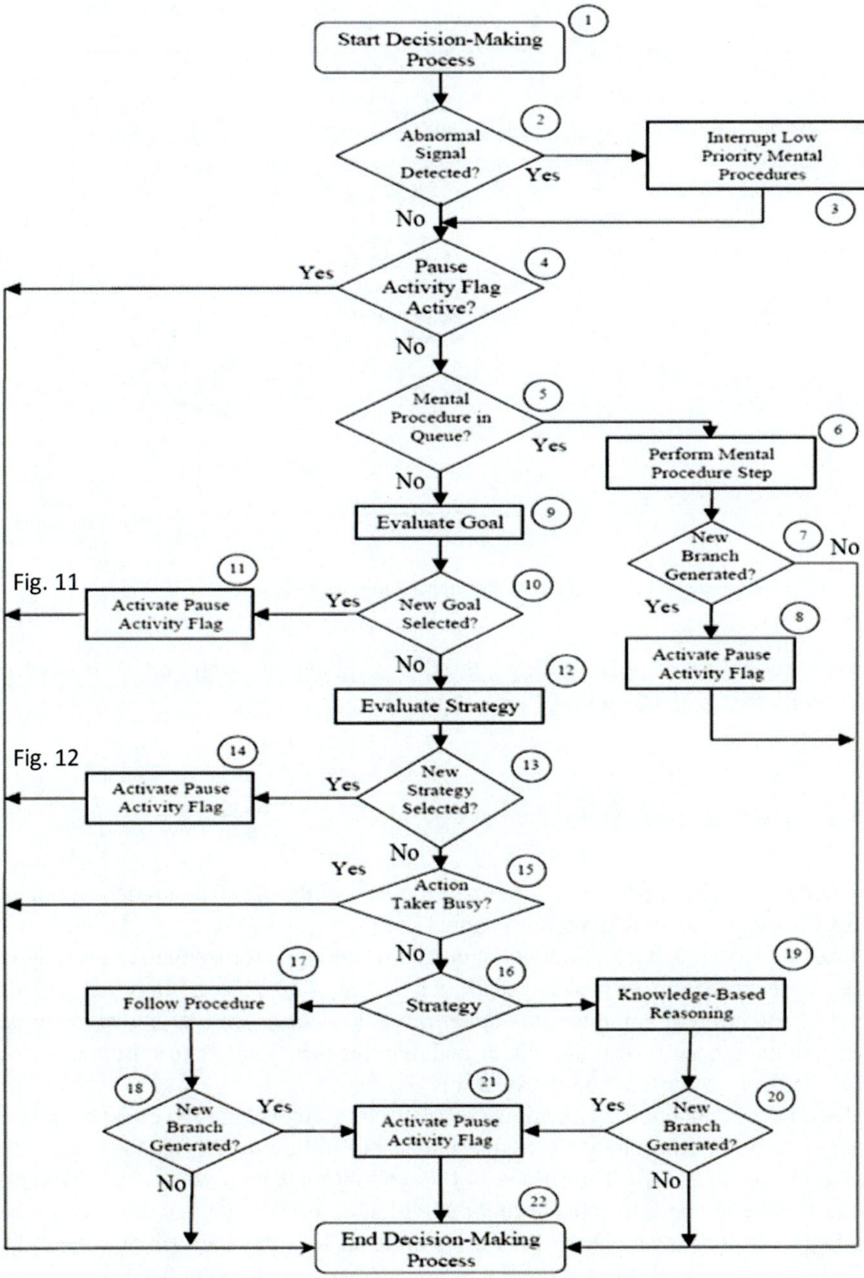

Fig. 5.10 Decision-making process in IDAC [34]

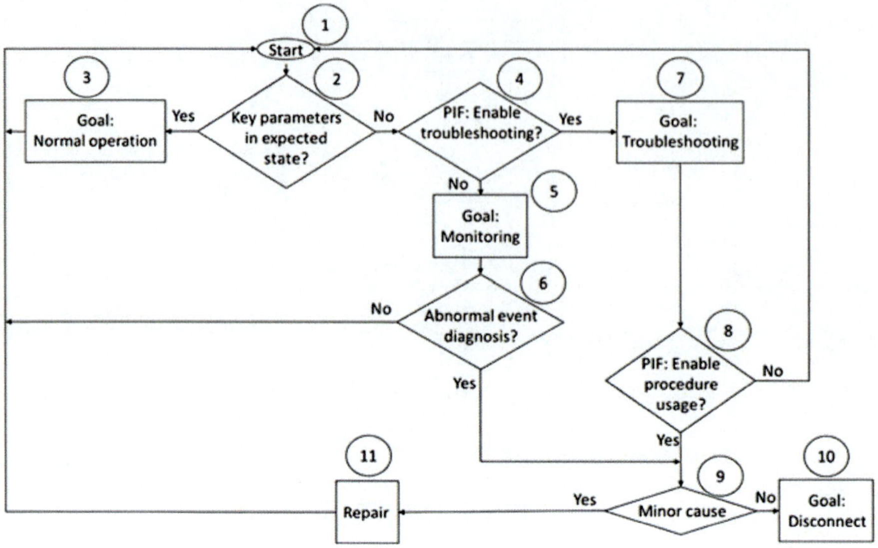

Fig. 5.11 Goal selection process, (PIF: performance influencing factors)

nodes have the most impact among other factors based on careful consideration of important aspects and factors.

5.5.2 Decision-Making Process

The decision-making process is the main module of the IDAC approach to model human behavior. Its flow diagram is shown in Fig. 5.10 [34].

The decision-making starts with an update to the operator's situation awareness based on new information observed (block 1, Fig. 5.10). Then abnormal signals should be identified. An abnormal signal refers to a situation that is incompatible with continued normal operation mode and requires the operators to shift their focus to an accident response mode (block 2, Fig. 5.10).

For a DP system onboard a drilling vessel, the abnormal condition could be failure of a sensor or entering the yellow zone/status. The yellow zone/status is defined as an actionable status [36]. The DP vessel is in yellow zone/status when one or several of the following conditions occur in the DP system: predefined performance limits reached, component or system failure resulting in loss of redundancy, or/and the vessel/rig maintains position although the vessel has lost its redundancy.

In this situation, the operator mental beliefs are "failure" and "normal operation." The "failure" mental belief is used to diagnose an abnormal condition; however, the "normal operation" belief is used in the process of goal selection. The analyst should assign suitable prerequisite conditions to all of these mental beliefs.

An abnormal signal is detected when the DP position or operating conditions exceed a preestablished threshold. The threshold value used for this is one of the static PSF factors included in the operator profile.

If an abnormal signal/alarm is detected, the operator will suspend implementation of any in-progress low priority activities that may have been previously initiated (block 3, Fig. 5.10), e.g., drilling is no longer performed once the DP has entered the yellow zone/status.

Whenever the operator receives an information illustrating that the activity related to the "pause" is completed, the appropriate "pause activity" flag is set to false to allow the decision-making process (block 4, Fig. 5.10).

Time required for communication is considered in this step (block 3, Fig. 5.10). This process can prioritize the execution rule- and skill-based actions over procedurally driven tasks. There should be a map that provides means to identify situations where operators might perform actions; they believe to be reasonable given their situation awareness (block 5, Fig. 5.10). Examples of such actions include restarting DP computer when it is slow or use redundant reference system in case of noisy signals.

For a DP system, prerequisites can be associated with DP's component states, alarms, active procedures in use, and an operator's mental beliefs.

Each mental belief may be connected with a mental procedure (block 6, Fig. 5.10). Mental procedures specify memorized rule- and skill-based actions typically executed by an operator without reference to a written procedure. In general, mental procedures cover alarm response procedures, diagnostic procedures, control procedures, and mitigation procedures [34].

In the ADS, different possible scenarios of system behavior are generated according to the system´s operating and environmental conditions. If any new scenario generates (block 7, Fig. 5.10), pause activity flag would be activated (block 8, Fig. 5.10). Then goals related to the new situation should be evaluated (block 9, Fig. 5.10).

The selection of a high-level goal is one of the first steps in the decision-making process. The operator reevaluates their high-level goal every time the decision-making process is activated. High-level goals could be categorized as monitor, maintain global safety, maintain normal operation, troubleshoot abnormal conditions, and avoid reaching point of no return.

The selection of a specific goal is based on information achieved by the operator and the status of the static performance influencing factors (PIFs) or performance shaping factors related to the goal selection. The flow diagram of goal selection in DP systems is presented in Fig. 5.11.

If any new goal is selected (block 10, Fig. 5.10), the pause activity flag would be activated (block 11, Fig. 5.10), i.e., if the goal should be updated, decision-making activities are suspended, and the new goal is sent to the communication blackboard for implementation during the next information processing phase (Fig. 5.7). Otherwise, the strategy would be assessed for the active goal (block 12, Fig. 5.10). Human operators use problem-solving strategy to perform their selected goal. Follow oral instruction, ask-for-advice, limited-reasoning, direct-matching, trial-and-error, wait and monitor, instinctive response, follow written procedures, and knowledge-based reasoning are examples of problem-solving strategies [34]. The flow diagram of the problem-solving strategy selection process is presented in Fig. 5.12.

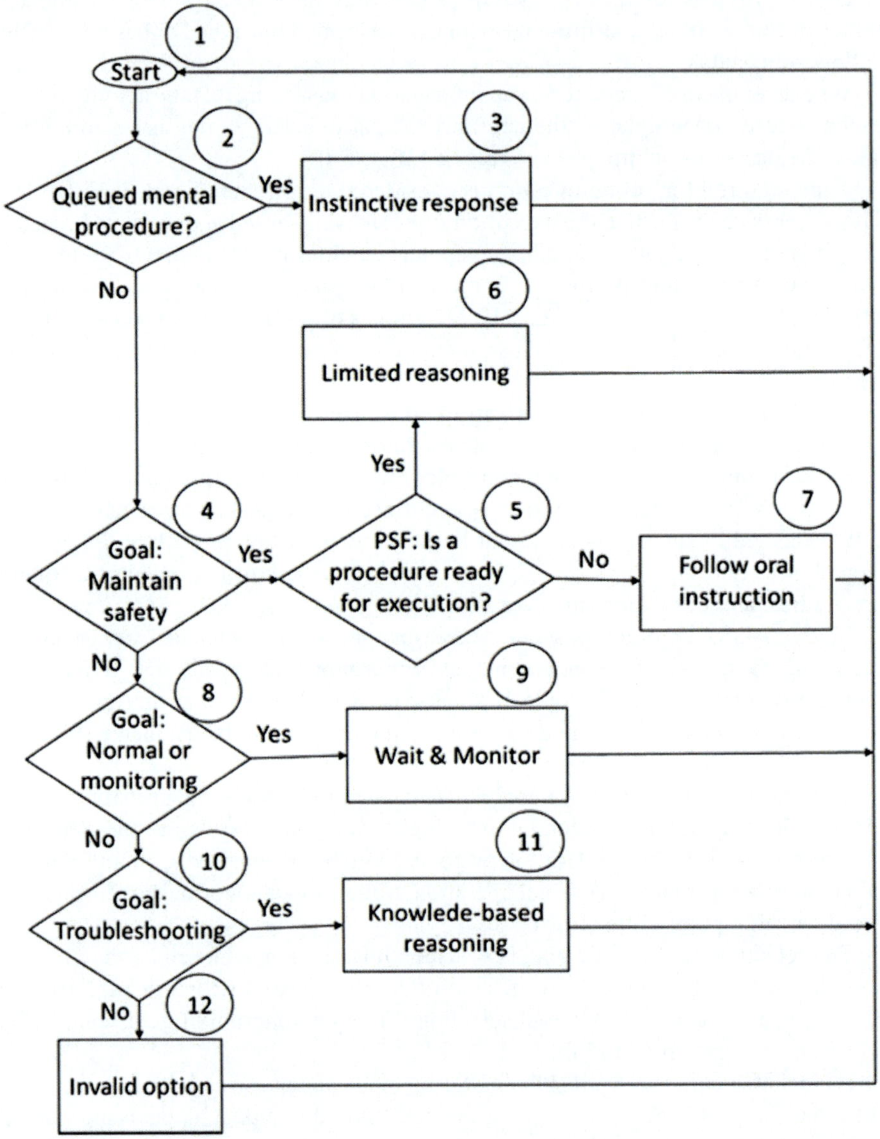

Fig. 5.12 Strategy selection process

If any new strategy is selected (block 13, Fig. 5.10), the pause activity flag will be activated (block 14, Fig. 5.10). If the strategy needs to be updated, decision-making activities are suspended, and the new strategy will be stored for implementation in the next information processing phase.

The action-taker performs any action that the decision-maker has made. If the action-taker is not able to receive and then execute an ordered action, further decision-making is suspended till the action-taker is available (block 15, Fig. 5.10).

If the action-taker is available to receive and execute an order from the decision-maker, an applicable problem-solving approach is followed according to the selected strategy (block 16, Fig. 5.10). Afterward, an appropriate procedure action or knowledge-based reasoning is identified according to the selected strategy (blocks17, 19, Fig. 5.10). Then, a new branch of appropriate procedure action or knowledge-based reasoning is generated accordingly (blocks 18, 20, Fig. 5.10). Pause activity flag is activated until all branches are generated (block 21, Fig. 5.10). These branches represent possible human behavior scenarios after an incident. Based on the branching points, event sequence diagrams of the human behavior could be developed and connected to the system event sequence diagram.

As mentioned before, the decision-making process could be simplified and presented as a Bayesian network [35]. It is assumed that operator(s) communicate and interact to understand the situation and to be able to make an appropriate decision in a short time. Finally, the recovery tasks are concluded by an operator based on the gathered information and time and operational constraints. For instance, three main decisions could be performed as recovery tasks in a DP drilling unit in short time [37].

1. Change position reference, i.e., deselect the faulty position reference and select an alternative position reference.
2. Recalibrate reference origin in DP control system.
3. Deselect faulty vessel sensor.

The failure probability of decision-making depends on multiple factors and could be calculated using the BN presented in Fig. 5.13. The presented BN in this figure is the Bayesian network of decision-making failure probability in DP systems. The highlighted red node has the most, and highlighted blue nodes have the least impact on

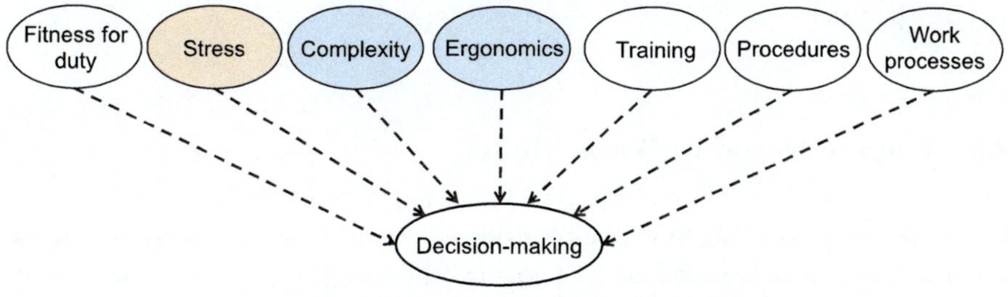

Fig. 5.13 Bayesian network of decision-making failure probability in DP systems

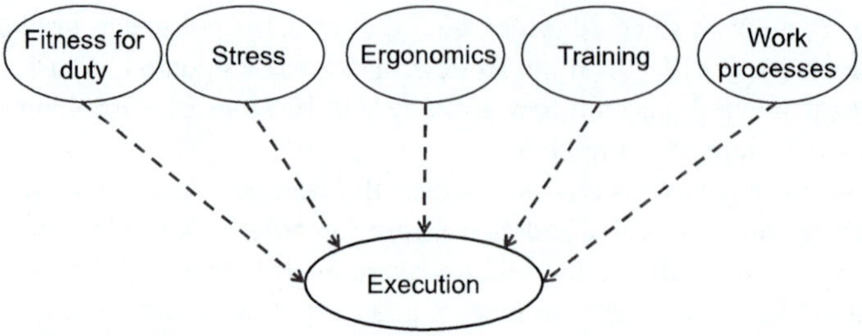

Fig. 5.14 Bayesian network of action execution failure probability

decision-making process based on careful consideration of important safety aspects and risk factors.

5.5.3 Action Execution Process

The final step of the IDAC cognitive model is action execution. In ADS-IDAC, action execution refers to active interactions with the system dynamic simulator (ADS) that changes the operating state of subcomponents, directed information gathering, and the activation of procedure steps. The type of action and behavior in which it is performed are dependent on the operator's PSFs and the action-response mode [34, 38].

The action execution module could provide different inputs to the system components. For instance, it adjusts or sets set points of control systems. This module could simulate realistically the significant human–machine interactions. All the presented steps could be simulated using different probabilistic models and evaluated human failure probabilities accordingly. The simplified method to model the action execution process is to develop a BN. In this method, we will not go through all the steps that should be followed for execution (Fig. 5.12), and we only consider the PSF that affect execution. Figure 5.14 presents a BN for the action execution process. The probability of execution failure depends on multiple PSFs, including fitness for duty, stress, ergonomics, training, and work processes, which are presented as parent nodes in the BN.

5.6 Human Response Time Model

In most research, response time is not considered or is takes as a constant value in the simulation process. However, in actual incidents, the decision-making process takes some time, and the probability of human error is highly dependent on the dedicated response time. So, having an accurate dedicated response time improves human error probability models. Human response time depends on several different factors, and various stochastic distributions are proposed to evaluate the estimated response time [39]. In this section, time distributions of detection, diagnosis, decision-making and execution processes are presented and discussed.

5.6.1 Detection

The response time (RT) of operator(s)' detection may follow the Gamma distribution [39]. The gamma distribution is a two-parameter category of continuous probability distributions. Equation (5.2) presents a general form of the gamma distribution. The gamma distribution can be parameterized in terms of a shape parameter α and an inverse scale parameter β = 1/θ.

$$f(x|\alpha, \beta, \xi) = \frac{(x-\xi)^{\beta-1}\exp(-\frac{x-\xi}{\alpha})}{\alpha^{\beta}\Gamma(\beta)} \tag{5.2}$$

In [39], it is shown that the best value for shape and scale parameters of a Gamma distribution to model detection response time are 2 and 400, respectively. Shape and scale value could be updated according to the detection process in the system under study and the mean value of detection response time. Figure 5.15 presents the shape of the gamma distribution with these parameters, and the mean value of 23 s, i.e., on average, the detection process takes 23 s.

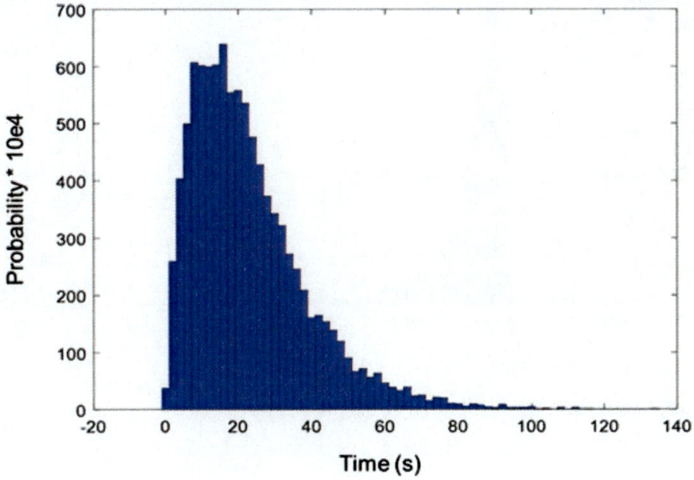

Fig. 5.15 Response time (RT) distribution of detection process

5.6.2 Diagnosis

The allocated time to the diagnosis process is stochastic. According to [40], the best distribution that fits human response time is the generalized inverse gamma distribution. In probability theory and statistics, the generalized inverse Gaussian distribution (GIG) is a three-parameter category of continuous probability distributions. The general form of the GIG distribution is presented in Eq. (5.3).

$$f(x) = \frac{(a/b)^{p/2}}{2K_p(\sqrt{ab})} x(p-1)e - (ax + b/x)/2, \quad x > 0, \tag{5.3}$$

where, Kp is a modified Bessel function of the second kind, a > 0, b > 0 and p are real parameters of the function. a, b, and p should be selected based on the system under study, nature of the diagnosis process and mean value of diagnosis response time. Herein, a, b, and p are considered to be equal to 1, 1, −0.5, respectively. These factors give a narrow tail that shows that there are some diagnosis processes that may take longer time. In addition, these factors give the mean value of the distribution equal to 20 s, which is the average required time for the diagnosis process, according [37] (Fig. 5.16).

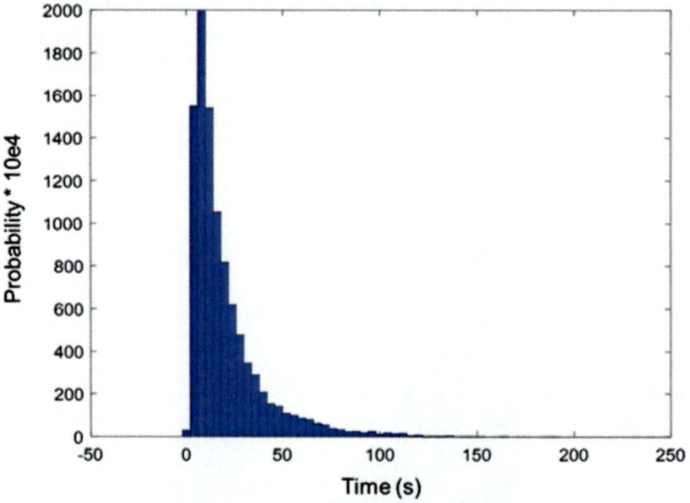

Fig. 5.16 Response time (RT) distribution of diagnosis process

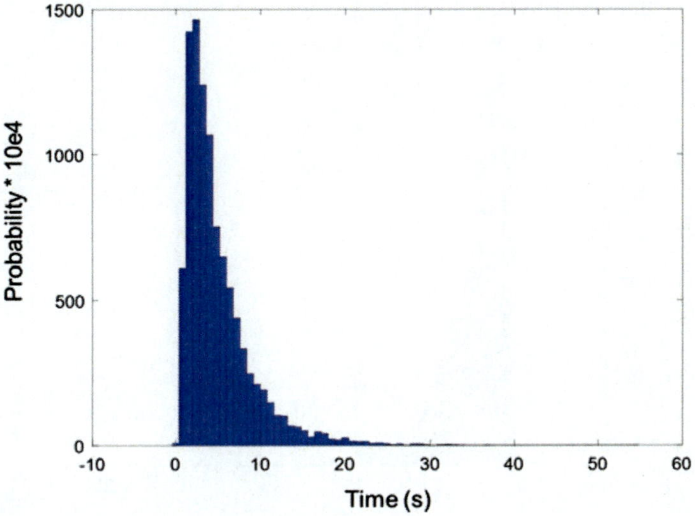

Fig. 5.17 Response time (RT) distribution of decision-making process

5.6.3 Decision-Making

The spent time at this step follows the GIG distribution, as in the diagnosis step. However, as the required time for this step is shorter than previous steps [37], a, b, and p parameters should be selected based on the nature of the decision-making and the mean value of the decision-making response time. Herein, a, b, and p parameters are considered equal to 2, 1, and 0, respectively, to achieve a mean value of 5 s. Figure 5.17 presents the GIG distribution for allocated time to the decision-making, with 5 s mean value [37].

5.6.4 Execution

The response time of execution follows the GIG distribution as in previous stages. The a, b, and p parameters should be selected according to the mean execution time of the system under study. Herein, a, b, and p parameters are considered equal to 2, 1, and 0, respectively, with 3 s mean value [35], i.e., on average, it takes 3 s to perform a task execution (Fig. 5.18).

Response time models estimate the allocated time to each step of the decision-making process including detection, diagnosis, decision-making and execution. The allocated time to each step affects the human error probability of that step. In Chap. 6, the application of response time models in human error and system risk level evaluations is presented and discussed.

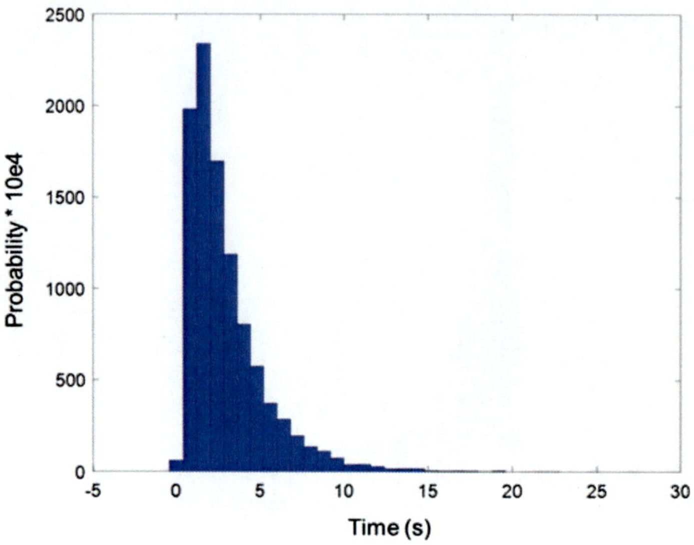

Fig. 5.18 Response time (RT) distribution of execution process

References

1. Hollnagel E (1998) Cognitive reliability and error analysis method (CREAM). Elsevier, Amsterdam, Netherlands
2. Cooper SE, Ramey-Smith AM, Wreathall J, Parry GW (1996) A technique for human error analysis (ATHEANA) (No. NUREG/CR--6350). Nucl Regul Comm
3. Kirwan B (1998) Human error identification techniques for risk assessment of high risk systems—Part 1: review and evaluation of techniques. Appl Ergon 29(3):157–177
4. Yoshida K, Yokobayashi M, Tanabe F, Kawase K (1996) Development of AI-based simulation system for man-machine system behavior in accidental situations of nuclear power plant. J Nucl Sci Technol 33(2):110–118
5. Smidts C, Shen SH, Mosleh A (1997) The IDA cognitive model for the analysis of nuclear power plant operator response under accident conditions. Part I: problem solving and decision-making model. Reliab Eng Syst Saf 55(1):5
6. Mosleh A, Chang YH (2004) Model-based human reliability analysis: prospects and requirements. Reliab Eng Syst Saf 83(2):241–253
7. Hu Y, Parhizkar T, Mosleh A (2021) Guided simulation for dynamic probabilistic risk assessment of complex systems: concept, method, and application. Reliab Eng Syst Saf. p 108047
8. Parhizkar T, Mosleh A (2022) Guided probabilistic simulation of complex systems toward rare and extreme events. The 68th Annual Reliability & Maintainability Symposium (RAMS), Tucson, Arizona
9. Hamed S.N, Parhizkar T, Mosleh A (2021) Simulation based probabilistic risk assessment (simpra): risk based design. 31st European Safety and Reliability Conference, 19–23 September 2021, Angers, France
10. Lund T (2007) Confidential incident investigation report, Loss of DP at XXX. TeekayMarine
11. Members/NBN T (2006) Significant incident investigation XXX DP incident on XXX. Confidential investigation report, Teekay Shipping (Canada) Ltd.
12. Oplenskedal A, Bjørheim LG (2019) Investigation of collision between Sjoborg supply ship and Statfjord A. Pet Saf Authority Norway

13. Drilling O (2011) Investigation report level 2, Rig excursion reached red watch cycle. Deapsea Stavanger
14. DNV (2011) Dynamic positioning systems-operation guidance
15. Flin R, Slaven G (1994) The selection and training of offshore installation managers for crisis management. HSE Books, Suffolk, England
16. Macdonald CJ, Weingarth L (2006) DPO assessment and improvement. Dynamic Positioning Conference, Houston, USA
17. Parhizkar T, Utne IB, Vinnem JE, Mosleh A (2021) Supervised dynamic probabilistic risk assessment of complex systems, part 2: Appli-cation to risk-informed decision making, practice and results. Reliab Eng Syst Saf 208:107392
18. Chang YHJ, Mosleh A (2007) Cognitive modeling and dynamic probabilistic simulation of operating crew response to complex system accidents. Part 4: IDAC causal model of operator problem-solving response. Reliab Eng Syst Saf
19. Hogenboom S, Parhizkar T, Vinnem JE (2021) Temporal decision-making factors in risk analyses of dynamic positioning operations. Reliab Eng Syst Saf 207:107347
20. Dos Santos IJL, França JE, Santos LFM, Haddad AN (2020) Allocation of performance shaping factors in the risk assessment of an offshore installation. J Loss Prevent Process Ind 104085
21. Bye A, Laumann K, Taylor C, Rasmussen M, Øie S, van de Merwe K, Øien K, Boring R, Paltrinieri N, Wærø I, Gould K (2017) The petro-HRA guideline
22. Sandhåland H, Oltedal H, Eid J (2015) Situation awareness in bridge operations–a study of collisions between attendant vessels and offshore facilities in the North Sea. Saf Sci 79:277–285
23. Isaac A, Shorrock ST, Kennedy R, Kirwan B, Andersen HB, Bove T (2003) Technical review of human performance models and taxonomies of human error in ATM (HERA) report, European Air Traffic Management Programme, Eur Organ Saf Air Navig
24. The International Marine Contractors Association (2005–2017) Dynamic positioning station keeping review, incidents and events reported for (2004–2015). International Marine Contractors Association
25. Chang YHJ, Mosleh A (2007) Cognitive modeling and dynamic probabilistic simulation of operating crew response to complex system accidents. Part 2: IDAC performance influencing factors model. Reliab Eng Syst Saf 92(8):1014–1040
26. Parhizkar T, Hogenboom S, Vinnem JE, Utne IB (2020) Data driven approach to risk management and decision support for dynamic positioning systems. Reliab Eng Syst Saf 201:106964
27. Li N, Huang J, Feng Y, Huang K, Cheng G (2020) A review of naturalistic decision-making and its applications to the future military. IEEE Access 8:38276–38284
28. Norwawi NM (2020) Recognition decision-making model using temporal data mining technique. J Inf Commun Technol 4:37–56
29. Nibbelink CW, Reed PG (2019) Deriving the practice-primed decision model from a naturalistic decision-making perspective for acute care nursing research. Appl Nurs Res 46:20–23
30. Saghiri AM, Meybodi MR (2016) An approach for designing cognitive engines in cognitive peer-to-peer networks. J Netw Comput Appl 70:17–40
31. Jalan A, Kumar N, Kannan P, Shashank MS (2019) U.S. Patent No. 10,268,779. U.S. Patent and Trademark Office, Washington, DC
32. Pan X, Lin Y, He C (2017) A review of cognitive models in human reliability analysis. Qual Reliab Eng Int 33(7):1299–1316
33. Chen W, An J, Li R, Li W (2017) Review on deep-learning-based cognitive computing. Acta Automat Sinica 43(11):1886
34. Coyne K (2009) A predictive model of nuclear power plant crew decision-making and performance in a dynamic simulation environment, Doctoral dissertation, Department of Mechanical Engineering, University of Maryland
35. Parhizkar T, Vinnem JE, Utne IB (2020) Dynamic probabilistic safety assessment framework to assist decision-making in complex systems, case study: dynamic positioning drilling unit. Ocean Engineering, 2020, Submitted

36. Cruz DF, Fonseca DDR (2017) WSOG and emergency disconnection guidelines. In: OTC Brazil. Offshore Technology Conference
37. Chen H, Moan T, Verhoeven H (2008) Safety of dynamic positioning operations on mobile offshore drilling units. Reliab Eng Syst Saf 93(7):1072–1090
38. Chang YHJ, Mosleh A (2007) Cognitive modeling and dynamic probabilistic simulation of operating crew response to complex system accidents: Part 3: IDAC operator response model. Reliab Eng Syst Saf 92(8):1041–1060
39. Palmer EM, Horowitz TS, Torralba A, Wolfe JM (2011) What are the shapes of response time distributions in visual search? J Exp Psychol Hum Percept Perform 37(1):58
40. Ma T, Holden JG, Serota RA (2016) Distribution of human response times. Complexity 21(6):61–69

Chapter 6
Dynamic Risk Assessment to Assist Decision-Making in Emergency Situations

An emergency is a situation that poses an immediate risk to health, life, property, or environment. Most emergencies require urgent intervention to prevent a worsening of the situation, although in some situations, mitigation may not be possible, and agents may only be able to offer alleviating care for the aftermath. Online risk assessment in emergencies should be performed in real-time, which means that dynamic probabilistic risk should be assessed in a very short execution time. The accuracy and execution time are highly dependent on the number of predicted scenarios. When predicting all possible scenarios, the number of scenarios could grow dramatically over time requiring long execution time. As presented in Chap. 2 (Sect. 2.5), optimization methods could assist in the scenario generation related to simulation based dynamic probabilistic risk assessment methods, enabling risk assessment in an acceptable execution time that meet emergencies requirements.

In this chapter, a dynamic probabilistic risk model in emergencies is proposed for dynamic positioning (DP) systems. In addition, in the following sections, the optimization method, discussed in Chap. 2 (Sect. 2.5), to generate failure scenarios in a very short execution time is presented for a case study.

6.1 Dynamic Probabilistic Risk Model of Decision-Making in Emergencies

Figure 6.1 presents the event flow diagram of decision-making in an emergency. In this figure, different stages of detection, diagnosis, decision-making, and execution are presented. After an initial event, operators detect the event. Then diagnosis would be performed, and based on that, decisions would be made and executed. This is a general process for decision-making in emergencies in a complex system [1, 2]. It is a simplified version of information, decision, and action in a crew context (IDAC) model, presented in Chap. 5 (Sect. 5.5). The first two columns include detection

T. Parhizkar et al., *Online Probabilistic Risk Assessment of Complex Marine Systems*, Springer Series in Reliability Engineering,
https://doi.org/10.1007/978-3-030-88098-9_6

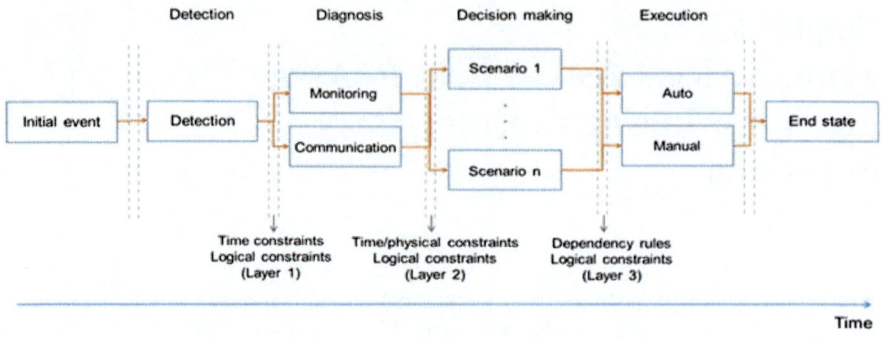

Fig. 6.1 Event flow diagram of decision-making in an emergency

and diagnosis process, which are the simplification of information filtering process presented in Chap. 5 (Sect. 5.5.1). The third and fourth columns are decision-making and execution, which are the simplification of decision-making process (Chap. 5, Sect. 5.5.2) and action execution process (Chap. 5, Sect. 5.5.3), respectively.

As presented in Fig. 6.1, each of these stages (columns) includes different events. The events in each stage should be adjusted based on the complex system functionality, e.g., the detection process in a dynamic positioning (DP) system is different from a nuclear power plant, and detection events should be specified accordingly. A specified event sequence diagram in emergencies for a DP system is presented in Fig. 6.2.

The first stage presents detection alternatives. Alarms represent the most common means of incident detection; however, sometimes operators perform other methods, including checking the vessel's position or making other visual detections, such as of the engine or weather status. The probability of alarm detection depends on an alarm system's health status, as well as an operator's performance shaping factors. Moreover, the occurrence probability of performing a "position check" or "other visual detection" is contingent on an operator's performance shaping factors (PSFs). For instance, it is more probable that an experienced operator performs successful "other visual detections" than an entry-level operator. The failure probabilities of these three detection processes are a function of an operator's performance shaping factors and can be calculated using Bayesian networks (BNs).

The next stage presented in the dynamic event sequence diagram (DESD) is the diagnosis phase. In this stage, operators try to figure out the status of the system and control the situation in order to return the system to its normal operation. In this stage, if there is enough time, operators check DP systems (propulsion, reference, computer, power, human error, and environmental factors) in order to identify the main causes of the incident as well as the consequences (drift-off, drive-off and potential incidents). Alternatively, operators attempt to control the situation using the available information from the detection stage and the DP screens. In addition to available time, the probability of performing diagnosis of components depends on detection accuracy. If the detection accuracy is high, component diagnosis is

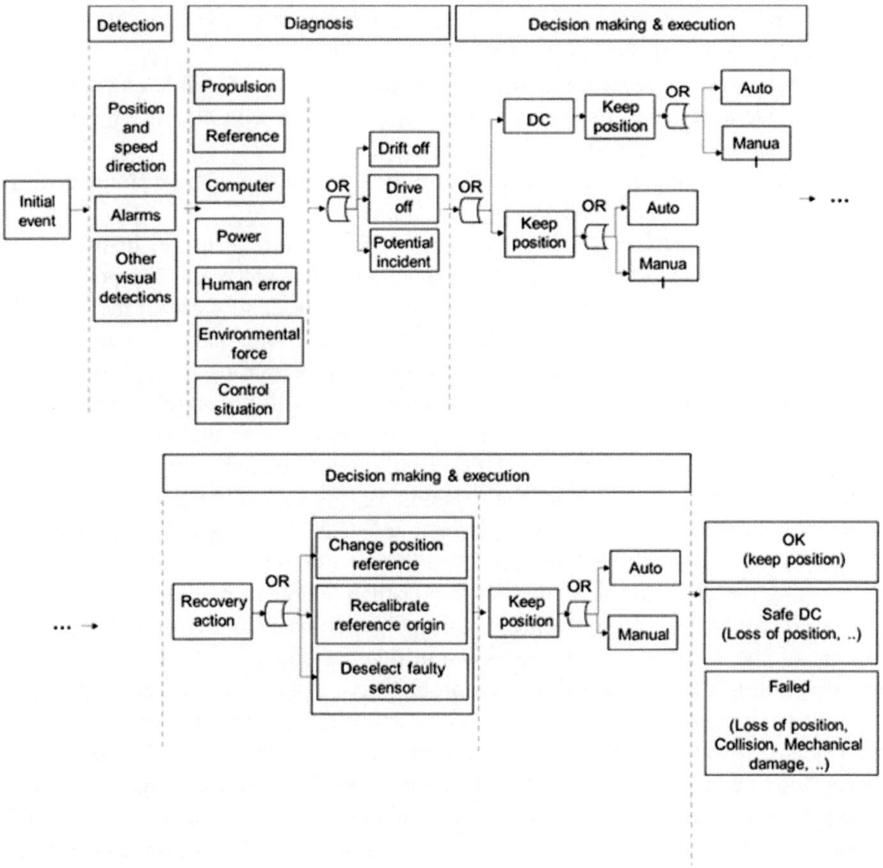

Fig. 6.2 Dynamic event sequence diagram in emergencies for dynamic positioning (DP) systems (DC: disconnection)[1]

unnecessary, as sufficient information has already been attained in the detection stage.

If there is enough time, operators will check the DP system, including propulsion, reference, computer, power, and human error,[2] and environmental factors. The failure probability of each of these subsystems/failure events can be evaluated using the associated fault trees (FTs). In addition, the probability of selecting a system to be diagnosed is assumed to be equal to the probability of not needing to diagnose, unless

[1] The end states are:

OK: DP system maintains position.

Safe DC: DP system loss its position (drift-off or drive-off), but it is safely disconnected.

Failed: DP system loss its position and disconnection are not performed properly, that results in collision or mechanical damage.

[2] Operator(s) would check if they have done a mistake.

a clear alarm from one system is detected. In such a case, the probability of selecting a related system to the alarm is higher.

Following system deviation detection and diagnosis, the possible consequences should be identified. In this step, the probabilities of the consequences, including drive-off, drift-off, potential incident, and dismiss alarm are calculated. If there is no fault in the components and or related to the environmental conditions, dismiss alarm is selected as an active event in the DESD.

The next stage of the DESD is decision-making and execution. In this stage, operators decide to stop the vessel (disconnection (DC)) or try to keep position. In either case, operators can select manual or automatic mode to perform the DC task. Selecting the task (disconnection or keep position) and mode (automatic or manual) depends on multiple factors, including standards, an operator's PSFs, components status, available time and so forth. For instance, in DP drilling vessels, well specific operating guidelines (WSOG) are used to define the actions that should be taken by a dynamic positioning operator (DPO) in the event of certain changes to the DP vessel's capability [3]. Whenever the vessel is passing the yellow region (defined in WSOG) and entering the red region (defined in WSOG), the operator must perform disconnection. As a result, if the vessel operates in a "red situation," the probability of selecting the disconnection mode is higher than keeping position. In "green and yellow situations," selecting the keep position branch has a higher probability.

One of the other stages in the decision-making and execution is performing recovery actions. It is assumed that the probability of performing a recovery action depends on the time available. In situations with limited time, no recovery actions are performed. However, if there is sufficient time, recovery actions can be performed. According to expert knowledge, recovery actions of DP systems in emergency situations are limited to seven scenarios, which are presented in Table 6.1. The required execution times of each recovery scenario are gathered based on expert knowledge as well and are shown in the final column of Table 6.1.

The failure probability of each event presented in the DESD (Fig. 6.2) can be calculated using the related fault trees and BN presented in Chap. 2, Sect. 2.3 [4–8].

Figure 6.3 presents the failure probability distributions of human error at different stages of detection, diagnosis, decision-making, and execution. A response time

Table 6.1 Recovery actions of each system and related timelines, based on expert judgments

Subsystem	Recovery actions by operator	Required execution time (s)
Reference	Change position reference	3–5
	Recalibrate reference origin	50–60
	Deselect faulty sensor	3–5
Control	Reference system recovery	3-60
	Tuning software	30
Power	Start new generator and connect it	30–60
Propulsion	Safe start of equipment on alternative switchboard	30–60

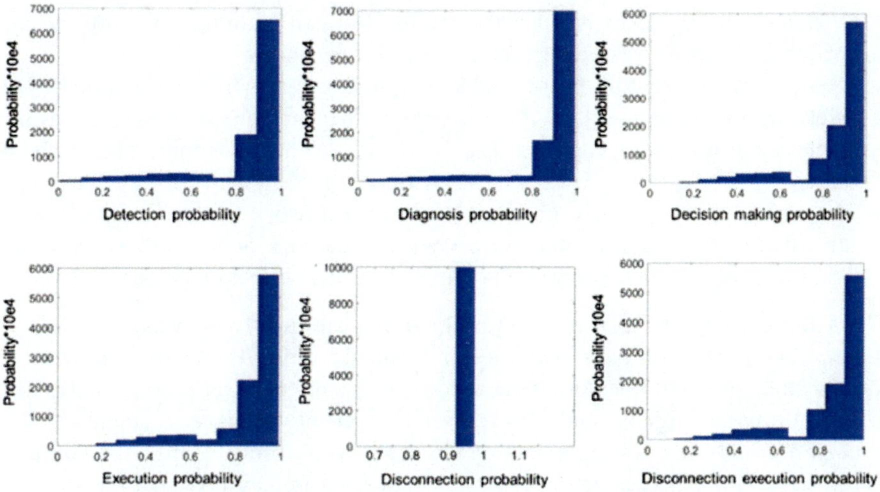

Fig. 6.3 Probability distribution of the successful decision-making process

model makes the human error model more accurate and reliable as we use a distribution for human error, instead of just a number. These failure distributions could be used in the quantification process of human error in risk assessment. As we have a failure distribution, we could use a sampling method such as the Monte Carlo[3] method to perform random sampling and provide human error quantities.

The failure probability of the decision-making process depends on the allocated time to this process. The time allocated to each event in the decision-making process (detection, diagnosis, decision-making, execution) is calculated using response time models presented in Chap. 5 (Sect. 5.6). The probability as a function of the allocated time could be calculated using the BNs presented in Chap. 5 (Sect. 5.5).

The distributions, presented in Fig. 6.3, are derived after selecting 10,000 random response times for each event. The random allocated times are selected from the distributions presented in Chap. 5 (Sect. 5.6). Then, these values are taken as inputs and the related probabilities are calculated based on the BNs presented in Chap. 5 (Sect. 5.5). In this model, it is assumed that:

- An incident has happened, and the yellow region has been entered.
- The complexity level of the task is normal, i.e., incident is not very complex and task complexity has neither a negative nor a positive effect on performance.
- Based on the operation and environmental conditions, the DP will enter the red region after 500 s, i.e., the available time for decision-making and recovery action to get back to the position is 500 s.
- Operator stress level is normal, i.e., operator(s) does not experience threat stress and the stress level has no negative effect on performance.

[3] Monte Carlo method is a class of computational algorithms that relies on repeated random sampling to obtain numerical results.

- Operator training level is normal, i.e., the level of training does not reduce performance nor to a large degree improve performance.
- Operator fitness for duty is acceptable, i.e., the operators have good attitudes to safety and work conduct and there is explicit management support to prioritize safety when that is appropriate. The operator(s) also shows mindfulness about safety.
- The teamwork is adequate on one or several teamwork factors that have been identified as important for the performance of the task or scenario in question. Teamwork has neither a negative nor a large positive effect on performance.

As can be seen in Fig. 6.3, event probabilities are mostly between 0.80 and 1. It should be noted that the probability of events occurring is independent, and as the available time before entering the red region is sufficient (on average, 500 s is a sufficient time to diagnose and take action), each event will have an adequate time to be performed. As a result, their accuracy (event probability) is high. In addition, as can be seen, the disconnection event has a constant probability of 0.94 in all runs. The reason is that auto-disconnection has a fixed allocated time; therefore, the related fault tree and BN result in a same value for all 10,000 runs.

As presented in Fig. 6.2, the end states of the emergency scenario are OK, disconnection (DC), and mechanical damage (MD). Figure 6.4 presents the occurrence probabilities of these end states. As is illustrated, the mode of the OK probability distribution is 0.95. It means that the probability of successful decision-making, action taking and get vessel back to a safe position, is 0.95. Disconnection refers to an appropriate disconnection without any damage to the system given that the red region is reached, and its probability is 0.05. MD is a failure in proper disconnection.

Figure 6.5 presents the allocated time distribution of all events in the decision-making and action taking process. The results are derived based on the 10,000 runs. As the number of runs increases, the pattern becomes more similar to the presented human RT distributions in Chap. 5, Sect. 5.6.

According to [9], there are some overlaps between the performed tasks in the decision-making process steps. The detection, diagnosis, and decision-making process could overlap at some points. The allocated time to overlaps of these processes are presented in [9] and based on those values the distribution of the overall allocated time from detecting an incident up to the end states is presented in

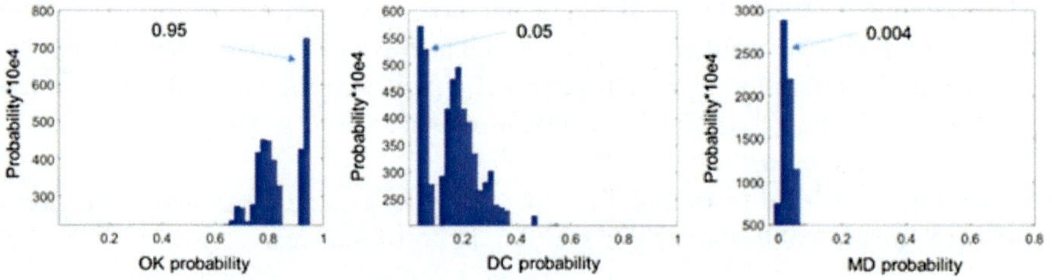

Fig. 6.4 End states probability distributions, (DC: disconnection; MD: mechanical damage)

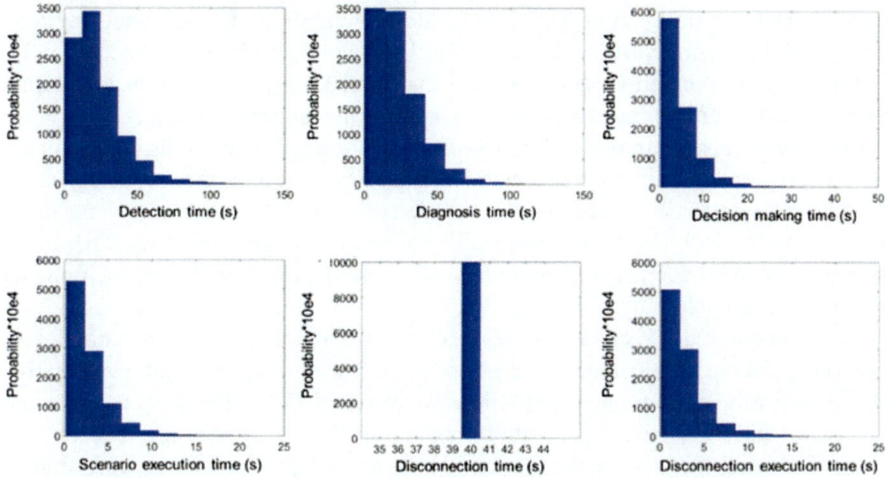

Fig. 6.5 Allocated/response time distribution for all decision-making process events

Fig. 6.6 Probability distribution of the overall allocated time from detection to execution

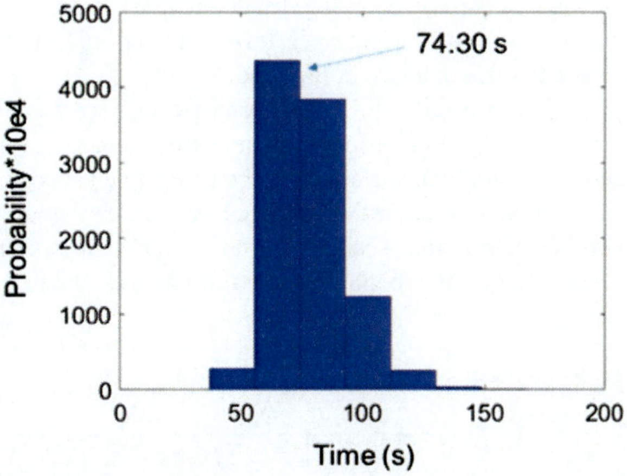

Fig. 6.6. The mode of the distribution is 74.30 s. In other words, the decision-making process from detection to taking action takes about 74 s that is shorter than 500 s.

6.2 Supervised Failure Scenario Generation in Emergencies

The system state, operational, and environmental variables affect the detection, diagnosis, decision-making, and execution process by the operator(s). In order to have a more relevant online risk model for decision support, dynamic event sequence diagram for decision-making (Fig. 6.2) should take inputs from system and environmental conditions in real-time. Then the risk level should be calculated accordingly. Considering the interactions between the decision-making process, the system status, operational, and environmental conditions increase the model accuracy. Such interactions, however, also increase the number of possible scenarios that could happen in the system, resulting in higher execution time. In the supervised failure scenario generation, the most important scenarios are generated (not all possible scenarios), meaning those scenarios with the highest probability of occurring are prioritized. Therefore, the execution time decreases significantly, while the accuracy of the model, remains within an acceptable range.

In this section, a DP system is presented as an example to show the complexity when considering human and system interactions. The number of possible alternative scenarios for the system operation/failure is presented. Then, the steps to apply the supervised failure scenario generation method on the DP system are presented in detail. The developed model in this section is further applied to different case studies which are presented in Chap. 7 (Sect. 7.2, 7.3, and 7.4).

In the DP example, it is assumed that the DP system has two states of failed and healthy; and three operational and environmental factors are considered affecting the vessel position, available time, and alarm status. All these factors, in addition to the system status, affect operators' decision-making process. Table 6.2 presents all considered parameters in the model.

In this example, multiple alternative scenarios could occur after an initial undesired event. In an emergency, by detecting an initial event, the states of the operational and environmental variables can be determined according to sensor data. System and human behavior operation modes, however, can assume any state. Consequently, the number of possible scenarios after an incident is equal to all combinations of the system and human behavior operation modes, which in this case is 5376.

Table 6.2 Parameters definition in the model

Component	Operation mode	State
k = 1 Operational and environmental variables	j = 1 Position	i = 1 Ok/i = 2 Lost
	j = 2 Available time	i = 1 Low/i = 2 High
	j = 3 Alarm	i = 1 On/i = 2 Off
k = 2 Human behavior	j = 1 Monitoring	i = 1 Done/i = 2 In progress
	j = 2 Detection types 1, 2, 3	i = 1 Done/i = 2 In progress
	j = 3 Disconnection types 1, 2	i = 1 Done/i = 2 In progress
	j = 4 Keep position types 1, 2	i = 1 Done/i = 2 In progress
	j = 5 Recovery action types 1–7 (Table 6.1)	i = 1 Done/i = 2 In progress
k = 3 System state	j = 1 Components	i = 1 Faulty/i = 2 Healthy

$$\begin{aligned}
&\text{Number of alternative scenarios at time t}\\
&= \text{Monitoring states} \times \text{Detectionstates}\\
&\times\ \text{Disconnectionstates} \times \text{Keep position states}\\
&\times\ \text{Recovery states} \times \text{System states}\\
&= ([\text{Operation mode: Monitoring}]\ \text{types} \times \text{States})\\
&\times\ ([\text{Operation mode: Detection}]\ \text{types} \times \text{States})\\
&\times\ ([\text{Operation mode: Disconnection}]\ \text{types} \times \text{States})\\
&\times\ ([\text{Operation mode: Keep position}]\ \text{types} \times \text{States})\\
&\times\ ([\text{Operation mode: Recovery}]\ \text{types} \times \text{States})\\
&\times\ ([\text{Operation mode: Components}]\ \text{types} \times \text{States})\\
&=\ (1 \times 2) \times (3 \times 2) \times (2 \times 2)\\
&\times\ (2 \times 2) \times (7 \times 2) \times (1 \times 2) = 5367
\end{aligned} \tag{6.1}$$

As time progresses, the number of alternative scenarios increases significantly and can be calculated using Eq. (6.2).

$$\text{Number of alternative scenarios} = (n)^k = 5376^k \tag{6.2}$$

The number of scenarios after k iteration is presented in Table 6.3.

It should be noted that not all these combinations are viable, and feasible connections should be defined based on the complex system (CS) operating characteristics, limitations, and study scope. Even by considering all possible and feasible scenarios, the execution time increases significantly, as numerous possible scenarios can be connected to multiple event trees (ETs)/event sequence diagrams (ESDs)/ fault trees (FTs)/ Bayesian networks (BNs). In order to evaluate the dynamic risk level in real-time, an optimization model is developed to generate desired scenarios more efficiently. The basics of this method are explained in Chap. 2, Sect. 2.5.

The objective function of the optimization model for the example DP system is:

$$\begin{aligned}
\text{Max} &\sum_{k=1}^{l}\sum_{j=1}^{m}\sum_{i=1}^{n}\left(p_{kji} \times x_{kji}\right) + \sum_{k=1}^{l}\sum_{j=1}^{m}\sum_{i=1}^{n}\left(p_{kji-kji} \times x_{kji-kji}\right)_t\\
&+ \sum_{k=1}^{l}\sum_{j=1}^{m}\sum_{i=1}^{n}\left(p_{kji-kji} \times x_{kji-kji}\right)_{t-(t+1)},
\end{aligned} \tag{6.3}$$

Table 6.3 Number of alternative scenarios at each iteration (time step)

Number of time intervals	1	2	3	4
Number of scenarios	5376	28e + 6	15e + 10	83e + 13

where the first term presents the occurrence probability of each operation mode, presented in Table 6.2. The second term presents the connection probability between operation modes at time t, and the last term presents connection probabilities between operation modes at times t and t + 1. Moreover, p_i presents the probability of state i for operation mode j and component k, n is the number of states, m is the number of operation modes, l is the number of components, and x_i is a binary variable {1,0} that indicates the existence and/or non-existence of operation mode i. Table 6.2 presents the i, j, and k of each component, operation mode, and state for the presented case study.

The first constraint of the case study is that only one active (x = 1) state exists at a time. For instance, an alarm on and off signal cannot occur simultaneously. This constraint is presented in Eq. (6.4).

$$\sum_{i=1}^{n}(x_{kji}) = 1 \quad , \quad \text{for } j = 1, 2, \ldots, m; \quad k = 1, 2, \ldots, l \tag{6.4}$$

The possible connections between component states could be determined based on the functionality and logic of the system operation. The connection between components at time t is presented in Table 6.4. Those values are considered in the second term of the objective function, presented in Eq. (6.3).

The dependency rules presented in Table 6.4 convey the probabilities, presented as connection probabilities. These probabilities depend on the component characteristics at time t. For instance, with faulty components and low available time, it is more logical to disconnect the vessel. Consequently, the components (faulty) and available time (low) have a high connection probability to the disconnection execution component.

Table 6.4 Component dependency rules at time t

Component (state) at time t	Activated component (state) at time t	Connection probability
Alarm (Off)	Keep position (In progress)	$(P_{132-242})_t$
Alarm (On)	Detection (In progress)	$(P_{131-222})_t$
Available time (High) Components (Faulty)	Recovery action execution (In progress)	$(P_{122-252})_t$ $(P_{311-252})_t$
Available time (Low) Components (Faulty)	Disconnection execution (In progress)	$(P_{121-232})_t$ $(P_{311-232})_t$
Alarm (On) Components (Healthy)	Keep position (In progress)	$(P_{131-242})_t$ $(P_{312-242})_t$
Position (Lost) Components (Healthy)	Keep position (In progress)	$(P_{112-242})_t$ $(P_{312-242})_t$
Position (Lost) Components (Faulty)	Disconnection execution (In progress)	$(P_{112-232})_t$ $(P_{311-232})_t$
Keep position (In progress)	Recovery (In progress)	$(P_{242-252})_t$

Table 6.5 Component dependency rules over time

State of component at time t	Activated component at time t + 1	Connection probability
Detection (Done)	Execution (keep position, disconnection, recovery actions)	$(P_{221-2j2})_{t-(t+1)}$
Execution (Done)	Monitoring	$(P_{2j1-212})_{t-(t+1)}$

In addition to connections at time t, components can be connected to other components at time t + 1. For instance, if the detection is completed, keep position, disconnection, or recovery actions should be performed. The dependency rules over time are presented in Table 6.5.

The probabilities presented in Table 6.5 are applied to the third term of the objective function, presented in Eq. (6.3).

An example of operating conditions over time is presented in Fig. 6.7. Generally, these values are sensor information from a DP system.

The operating conditions presented in Fig. 6.7 are x values, presented in Eq. (6.3), related to the operation modes at each time interval. For instance, at the first-time interval:

$$x_{131} = 1 \text{ (Alarm is On)} \tag{6.5}$$

$$x_{111} = 1 \text{ (Position is OK)} \tag{6.6}$$

$$X_{122} = 1 \text{ (Available time is High)} \tag{6.7}$$

According to Eq. (6.4):

$$X_{132} = 0 \text{ (Alarm is Off)} \tag{6.8}$$

$$X_{112} = 0 \text{ (Position is Lost)} \tag{6.9}$$

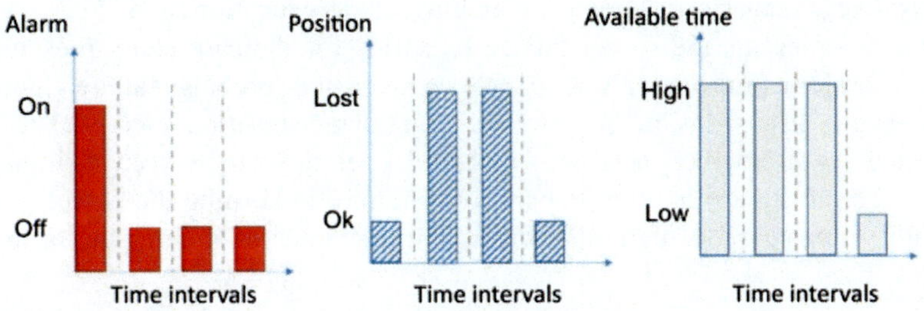

Fig. 6.7 Operating conditions over four-time intervals for the example

Table 6.6 Possible scenarios at each time interval

Time interval 1	Time interval 2	Time interval 3	Time interval 4
Detection (3)	Keep position (2) Detection (3) Disconnection (2)	Keep position (2) Detection (3) Recovery (7) Disconnection (2)	Keep position (2)

$$x_{121} = 0\,(\text{Available time is Low}) \tag{6.10}$$

At each time interval, x and p are generated according to Tables 6.4, 6.5, and Fig. 6.7. As shown in Fig. 2.7, at the first level, the alarm is on, the position is "OK," and the time is sufficient. According to Table 6.4 (second row), at this time interval, detection would be activated. At the next time step, the outcomes of the first-time interval are used as inputs; in addition, according to the dynamic simulation, the alarm is off, the position is "lost," and the available time is high. According to this information, the possible activated operation modes are generated accordingly. All possible operation modes for all the time intervals are presented in Table 6.6. The number of alternative scenarios of each operation mode (presented in Table 6.2) is shown in parentheses.

According to Table 6.6, the numbers of possible scenarios corresponding to each time interval in Fig. 6.7 are 3, 7, 14, and 2. The number of all combinations of these scenarios in this duration is 588, which is considerably less than the initial 5376, due to the dependency connections.

As time progresses in simulation, the number of scenarios increases significantly. In all the conventional methods, the probability of all the generated scenarios should be evaluated, which could result in a long execution time. However, in the proposed concept, an optimization algorithm is applied. In the optimization process, optimal x values for all operation modes are generated to obtain the most probable failure scenario. The x values determine the state of all operation modes of the most probable scenario. After defining the most probable scenario, the optimization model is executed again to obtain the second most probable scenario and corresponding x values are generated. The process is repeated until all failure scenarios are generated. It is noteworthy that in this method, failure scenarios are generated and sorted without solving all possible scenarios. Consequently, the "optimal" scenarios, i.e., the most likely scenarios, can be generated in a short execution time.

Table 6.7 presents the sorted failure scenarios for the case study presented in Fig. 6.7 after four-time intervals. As can be seen, the most probable failure scenario is that after auto-disconnection[4] a recovery action (change position reference) has been performed. As the recovery action is successful, operator(s) try to keep position. This failure scenario shows that the probability of failure in keeping the vessel position manually, in this case, is high (all scenarios at time interval 4 are failure of manual

[4] There is some time lag from the start of the auto-disconnection to completion where operators can act to avoid it.

Table 6.7 Failure scenarios after four-time intervals

No	Time interval 1	Time interval 2	Time interval 3	Time interval 4
1	Alarm detected	Auto-disconnection (DC)	Change position reference	Failure of manual kp[a]
2	Alarm detected	Auto DC	Recalibrate reference origin	Failure of manual kp[a]
3	Alarm detected	Auto DC	Deselect faulty sensor	Failure of manual kp[a]
4	Alarm detected	Auto DC	Reference system recovery	Failure of manual kp[a]
5	Alarm detected	Auto kp[a]	Change position reference	Failure of manual kp[a]
6	Alarm detected	Auto kp[a]	Recalibrate reference origin	Failure of manual kp[a]
7	Alarm detected	Auto kp[a]	Deselect faulty sensor	Failure of manual kp[a]
8	Alarm detected	Auto kp[a]	Reference system recovery	Failure of manual kp[a]
9	Position checked properly	Auto DC	Change position reference	Failure of manual kp[a]
10	Position checked properly	Auto DC	Recalibrate reference origin	Failure of manual kp[a]

[a] kp refers to keep position

keeping position). The rest of the scenarios presented in the table are the sorted from the more probable failure scenarios of the DP system, considering the presented environmental and operational conditions in Fig. 6.7.

Table 6.8 presents the execution time of two conventional, discrete dynamic event sequence diagram (Chap. 2, Sect. 2.4.1) and optimization-based (Chap. 2, Sect. 2.5.2; Chapter 6, Sect. 6.2) methods for different time intervals. As shown, at lower time intervals, the execution time of the conventional methods is lower. That is primarily because of the lower number of possible scenarios. As the number of possible scenarios increases, the execution time of the conventional method (dynamic event sequence diagram) increases significantly. However, the execution time of the

Table 6.8 Number of possible alternative scenarios and execution time at each time interval

	K = 1	K = 2	K = 3	K = 4	K = 5	K = 6
	3	21	294	588	8232	16,464
Conventional method	<10 s	<10 s	~10 s	~1 min	~2 min	~5 min
Optimization based method	10 s < t < 1 min					

optimization method increases only slightly. In this example, the modified particle swarm optimization (PSO)[5] is used as the PSO is a powerful algorithm for nonlinear models with discrete and continuous variables. However, it is noteworthy that other solution algorithms would result in almost the same execution time owing to the simplicity of the presented example. In more complex systems, an efficient solution algorithm should be selected based on the model characteristics and algorithm features.

As presented in Tables 6.7 and 6.8, the supervised failure scenario generation method is able to generate more probable failure scenarios in a much lower execution time compared to the conventional dynamic event sequence diagram method without any scenario prioritization. In [10], a more detailed example of the application of this method is presented, and results are discussed.

References

1. Hogenboom S, Parhizkar T, Vinnem JE (2021) Temporal decision-making factors in risk analyses of dynamic positioning operations. Reliab Eng Syst Saf 207:107347
2. Parhizkar T, Utne IB, Vinnem JE, Mosleh A (2021) Dynamic probabilistic risk assessment of decision-making in emergencies for complex systems, case study: Dynamic positioning drilling unit. Ocean Eng 237:109653
3. Cruz DF, Fonseca DDR (2017, Oct) WSOG and emergency disconnection guidelines. In OTC Brasil. Offshore Technology Conference
4. Parhizkar T, Hogenboom S, Vinnem JE, Utne IB (2020). Data driven approach to risk management and decision support for dynamic positioning systems. Reliab Eng Syst Saf 201:106964
5. Parhizkar T, Vinnem JE, Utne IB, Mosleh A (2020) Supervised dynamic probabilistic risk assessment of complex systems, Part 1: General overview. Reliab Eng Syst Saf 107406
6. Hu Y, Parhizkar T, Mosleh A (2021) Guided simulation for dynamic probabilistic risk assessment of complex systems: concept, method, and application. Reliab Eng Syst Saf 108047
7. Parhizkar T, Mosleh A (2022) Guided probabilistic simulation of complex systems toward rare and extreme events. The 68th annual reliability & maintainability symposium (RAMS), Tucson, Arizona
8. Hamed SN, Parhizkar T, Mosleh A (2021) Simulation based probabilistic risk assessment (simpra): risk based design. 31st European Safety and Reliability Conference, 19-23 September 2021, Angers, France
9. Bye A, Laumann K, Taylor C, Rasmussen M, Øie S, van de Merwe K, Gould K (2017) The petro-HRA guideline
10. Parhizkar T, Utne I, Vinnem JE, Mosleh A (2020) Supervised probabilistic risk assessment of complex systems, Part 2: Application to risk-informed decision making, practice and results. Reliab Eng Syst Saf 208:107392

[5] PSO is a population-based optimization algorithm inspired by the social behavior of fish schooling and bird flocking. The PSO algorithm shares many similarities with other evolutionary optimization algorithms, such as genetic algorithms. A main advantage of the PSO algorithm over other evolutionary methods is that only a few parameters need to be adjusted in the PSO algorithm.

Chapter 7
Case Studies

7.1 Application 1: Loss of Position of a Mobile Offshore Drilling Unit

On April 2018 at 12.38, a mobile offshore drilling unit (MODU) lost position during a workover of the riser and installation of a float tree on the well A-03 [1]. Several high waves push the MODU off position, that combined with the drilling operation that is using up the power, there is insufficient thrust available to keep the MODU on position and a drift-off occurs. Between the first wave and the 12-m excursion, there is not much time; there is maximum a minute at best to respond for the dynamic positioning operator (DPO).

The course of events is presented in Fig. 7.1. During this incident, the MODU passed the limits for the advisory, yellow, and the red state. The well specific operating guidelines (WSOG) for the studied MODU specifies that for the red state, it is at the offshore installation manager (OIM)'s discretion to decide if the situation is a threat to equipment and how to act. In this specific case, the OIM saw that the speed at which the MODU was moving off position was decreasing and the OIM expected the vessel to be able to regain position fast. Therefore, the OIM decided not to disconnect and move to the safe zone but rather inhibit the automatic disconnect and wait for the MODU to move back into position with the DP system. The vessel regained position after having a maximum excursion of 12 m.

The events here are analyzed using the standardized plant analysis risk-human reliability analysis (SPAR-H) method [2]. In addition to the scenario described in the incident report (scenario1), an alternative scenario is also analyzed (Scenario 2). Scenario 2 is similar to scenario 1; several high waves push the MODU off position, that combined with the drilling operation that is using up power, there is insufficient thrust to keep the MODU in position and a drift-off occurs. In the alternative scenario, it is assumed that the OIM is not on the bridge and that the dynamic positioning operator (DPO) is responsible for the decision-making and

T. Parhizkar et al., *Online Probabilistic Risk Assessment of Complex Marine Systems*, Springer Series in Reliability Engineering,
https://doi.org/10.1007/978-3-030-88098-9_7

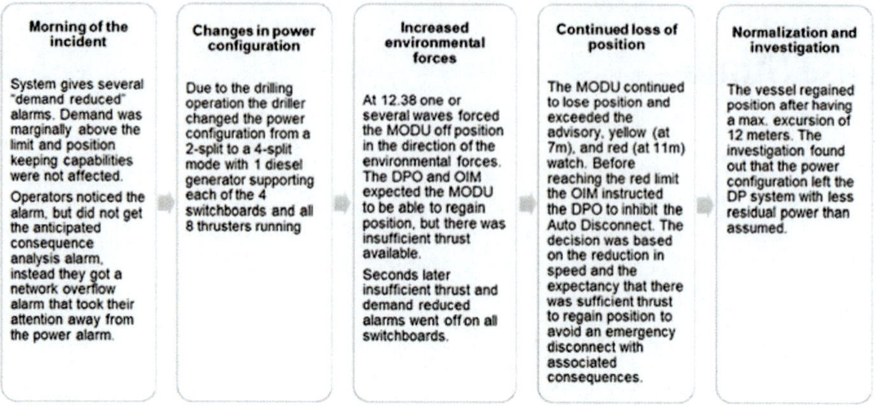

Fig. 7.1 Course of events of the mobile offshore drilling unit (MODU) incident based on investigation report [1], (DPO: dynamic positioning operator, OIM: offshore installation manager)

handling of the loss of position. The scenario has been included to present a realistic alternative in a decision-making situation and how this may affect the risk picture.

The human failure event for these two scenarios has been defined as: Failure to handle a drift-off. The performance shaping factors (PSFs) of the two scenarios are presented in Table 7.1 (codes in the table are explained in the text below the table).

Table 7.1 Performance shaping factors (PSFs) of the mobile offshore drilling unit (MODU) incident and alternative scenario without the offshore installation manager (OIM) on the bridge

PSFs	PSF levels	Evaluation of PSFs of incident with OIM on bridge scenario 1	Evaluation of PSFs of incident without OIM on bridge scenario 2
Available time	Nominal	A	B
Stress	High	–	C
	Nominal	D	–
Complexity	Nominal	E	F
Training	High	G	–
	Nominal	–	H
Procedures	Available, but poor	–	I
	Nominal	J	–
Ergonomics	Nominal	K	L
Fitness for duty	Insufficient info	–	M
	Nominal	N	–
Work processes	Insufficient info	–	O
	Nominal	P	–

The parameters in the table are, [3, 4]:

(a) Based on the information available from the incident, there was sufficient time available to perform the action, and therefore, available time is considered to be nominal.
(b) It is assumed that there is sufficient time, i.e., nominal, for the DPO to act, based on the time available during the incident with the OIM on bridge.
(c) Without the OIM on the bridge, the DPO will have to make a decision herself/himself. As the situation is described in the incident, an automatic or manual emergency disconnect would have happened, if the functionality was not inhibited. This could have potential safety and financial consequences.
(d) In the incident, the OIM had a good overview of the situation and said to have the situation under control. The PSF stress is therefore evaluated to be nominal.
(e) The description of the incident does not indicate that the diagnosis is difficult, but it is also not obvious. Therefore, complexity is evaluated to have a nominal effect on the human error probability (HEP).
(f) The description of the incident does not indicate that the diagnosis is difficult, but it is also not obvious. Therefore, complexity was evaluated to have a nominal effect on the HEP.
(g) The OIM is considered to have extensive experience with DP. Training and experience are therefore evaluated as high.
(h) Since this scenario is hypothetical, no assumptions are made on the level of experience/training of the DPO.
(i) The lack of descriptiveness in this scenario is assumed to lead to a difference in approach by the DPO contra the OIM. The procedures state that it is up to the discretion of the OIM to decide how to handle such a situation. If the OIM is not available, as is assumed for this scenario. The DPO is assumed to decide to disconnect and move to a safe zone to stabilize the power distribution.
(j) Procedures are available and are relatively simplistic and straightforward. However, they are not very descriptive. The effect of procedures is therefore considered to be nominal.
(k) No information available on the state of the ergonomics/human–machine interface (HMI).
(l) No information available on the state of the ergonomics/HMI.
(m) Since this situation is hypothetical, no assumptions are made on the fitness for duty of the DPO.
(n) The OIM is able to carry out tasks and there is no description of performance degradation. The PSF is therefore considered nominal.
(o) There is insufficient information available in the incident description to infer how work processes would affect the performance of the DPO.
(p) From the description of the incident, the decision-making seems quite unilaterally performed by the OIM. So, work processes are not considered a performance driver.

The difference between Scenario 1 and 2 is the manning on the bridge. During Scenario 1, the OIM and the DPO are on the bridge, as in the incident on the MODU. Scenario 2, however, assumes that the OIM is not able to come to the bridge in time and that the DPO has to handle the drift-off alone.

The Bayesian network (BN) of human errors is developed based on the model presented in Chap. 5, Subsect. 5.4, and input data comes from Chap. 3, Subsect. 3.3 [5, 6]. Table 7.2 presents the estimated human error probabilities in the two scenarios.

It can be observed that the slip and mistake probabilities increase significantly from scenario 1 to Scenario 2. This increase is largest in the rule-based mistake category due to the significance of the training factor.

System failure probability is modeled based on the fault tree (FT)/BNs presented in Chap. 2, Subsect. 2.3, for DP systems [7, 8]. It is assumed that failure of each of the main systems (thruster, power, reference and control systems) could result in the system loss of position (LOP). By calculating the failure probabilities of the components based on the data and fault trees presented in Chap. 2 and Appendix A, the DP system LOP probability is derived. Table 7.3 shows the probabilities of the end states occurrence in the event sequence diagram (ESD). LOP1, LOP2, LOP3, and LOP4 present loss of position due to propulsion, reference, power, and control system faults, respectively. As can be seen from Table 7.3, there is no difference between the probability of failure under manual and automatic operations for the first three end states (LOP 1–3). This lack of difference is due to the power, thruster and reference systems being independent of the way that the systems are controlled. However, the

Table 7.2 Human error probabilities in different systems for two scenarios

	State	Scenario 1	Scenario 2
Slip	Yes	0.0022	0.0222
	No	0.9978	0.9778
Mistake rule-based	Yes	0.0110	0.1011
	No	0.9890	0.8989
Mistake knowledge-based	Yes	0.0110	0.0511
	No	0.9890	0.9489
Human error in power sys	Healthy	0.9921	0.9432
	Faulty	0.0079	0.0568
Human error in ref sys	Healthy	0.9953	0.9641
	Faulty	0.0047	0.0359
Human error in thruster sys	Healthy	0.9981	0.9819
	Faulty	0.0019	0.0181
Manual control error	No error	0.9779	0.8358
	Error	0.0221	0.1642
Automatic control sys	Healthy	0.9916	0.9430
	Faulty	0.0084	0.0570

Table 7.3 End state probabilities in manual and automatic disconnection modes for the two scenarios

	Scenario 1		Scenario 2	
	Auto	Manual	Auto	Manual
LOP1[a]	0.0020	0.0020	0.0181	0.0181
LOP2[a]	0.0042	0.0042	0.0315	0.0315
LOP3[a]	0.0056	0.0056	0.0391	0.0391
LOP4[a]	0.0040	0.0218	0.0271	0.1496
OK	0.9843	0.9664	0.8842	0.7616

[a] LOP refers to loss of position

Table 7.4 Unavailability of dynamic positioning systems for the two scenarios

	Scenario 1	Scenario 2
Thruster system	0.0020	0.0181
Reference system	0.0048	0.0359
Power system	0.0079	0.0568
Automatic control system	0.0084	0.0570
Manual control system	0.0221	0.1642

fourth end state (LOP 4) presents the control system fault, and the probability of this state is affected highly by the method of control (manual or automatic).[1]

As presented in Table 7.3, the significance of the human factor can be observed in LOP4 where the probabilities of failure increase dramatically from the automatic to manual disconnection mode in both scenarios. The increased probability in the manual mode is more significant under scenario 2, due to lower manning where the human factor is more consequential due to higher stress level and lower level of training and procedure.

Comparison of the two scenarios indicates the importance of training since the probability of failure in the manual mode under Scenario 1 is even lower than the probability failure in the automatic mode under Scenario 2.

The unavailability U of systems/components could be calculated based on their failure probabilities. Table 7.4 presents the unavailability of thruster, reference, power, and control systems of the DP system under study. These values present the proportion of time that these systems are not in a functioning condition and are calculated from Eq. (7.1).

$$U = \frac{\text{Downtime}}{\text{Totaloperatingtime}} = \frac{MDT}{MDT + MTTF} \tag{7.1}$$

where U is the average unavailability, MDT is the mean downtime and MTTF is the mean time to failure. MDT and MTTF are functions of systems failure probability.

[1] Manual mode: operators use a joystick to control the position and heading of the DP system.

Systems failure probabilities have been calculated using FTs/BNs presented Chap. 2, Subsect. 2.3, and failure probabilities are derived from [9]. The unavailability values are the same for both manual and automatic modes for the thruster, reference, and power systems, but different for the control system. All values are significantly higher under Scenario 2 due to the human error performance shaping factors (PSF) and the effect of training and procedure in Scenario 1. The biggest increase in unavailability can be found between Scenario 1 to 2 for the manual control system where the human factor is more critical.

The results from Tables 7.3 and 7.4 could serve as input to online risk assessment and improved decision support. The unavailability of the different DP systems is part of the risk levels associated with staying on location. The results also indicate how to move off location in long term (not emergencies), if necessary, by using the unavailability of the manual and automatic control systems as input.

7.2 Application 2: Collision of a DP Shuttle Tanker

On November 13, 2006, a DP shuttle tanker collided with a floating storage and offloading unit (FPSO) as it started its loading operation. The shuttle tanker sustained some damage above the waterline in the bow area. The FPSO suffered damage to its stern and to some equipment in the after part of the poop deck. According to the investigation report, the shuttle tanker was operating in the DP automatic positioning mode and had just received and secured the mooring hawser. As the bow operator was in the process of preparing the loading hose transfer, a blackout of the starboard main switch board (MSWB) occurred. The blackout caused a loss of the starboard main propulsion as well as a loss of side thrusters' numbers 2 and 4, as these were powered by the said MSWB.

Unexpectedly, side thrusters' numbers 1 and 3 also stopped due to a lack of power. With only the port main propeller and rudder available on the DP, the vessel remained in DP automatic positioning mode. The DP system was not capable of maintaining or controlling its position with only one rudder and one propeller. About four minutes later, the shuttle tanker hits the FPSO with approximately 2.4 knots speed [10]. The impact caused visual damage to both vessels, but no physical injuries to personnel.

This application is modeled based on the cognitive concept presented in Chap. 5, Subsect. 5.5. Human behavior consists of three main steps of detection, diagnosis, decision-making, and execution. Figure 7.2 presents the possible events at each step of the event sequence diagram. The sequence of events, in the presented case study, is highlighted (gray boxes) in Fig. 7.2.

In this application, it is assumed that we only have the initial incident and ambient conditions, and we would like to generate more probable failure scenarios as the model output. More probable failure scenarios would help operators to make better decisions in emergencies.

To model this case study, the events in Fig. 7.2 are quantified based on the BN models presented in Chap. 5, Subsect. 5.5. The optimization-based DPRA model,

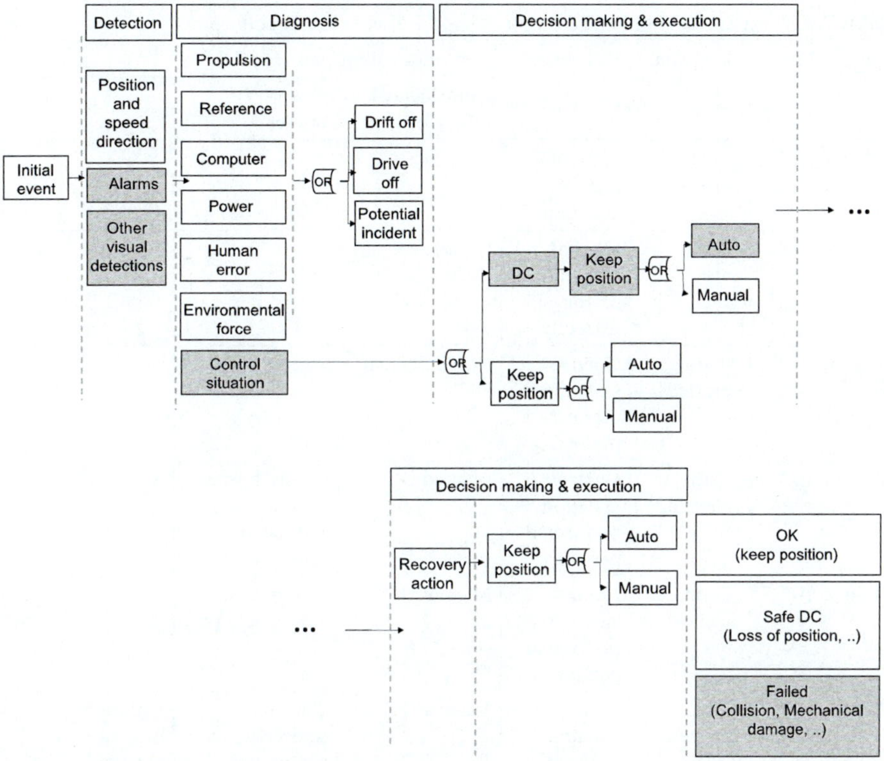

Fig. 7.2 Event sequence diagram of the DP shuttle tanker collision, gray blocks represent the event sequence diagram of the accident, (DC: disconnection)

presented in Chap. 2, Subsect. 2.5.2, is applied to the case study, and failure scenarios are generated accordingly. The inputs of the model are derived from the investigation report of this incident [10] and are presented as follows:

- Blackout has occurred.
- The operator has enough fitness for duty and training, and the stress level is normal.
- There is not enough time, and the DP vessel is in an emergency situation.
- The environmental conditions, including wind and waves (force and directions) are in normal conditions.
- The reference and computer systems work properly.
- The propulsion system including thrusters has failed.

Based on these inputs, the model is run, and all failure scenarios (271 scenarios) are generated [7, 8, 11–13]. The probability of system failure is equal to the sum of all failure scenarios probabilities, which is 0.726. The generated scenarios are sorted based on the failure probability, and the most probable scenarios are presented in Table 7.5.

Table 7.5 Most probable failure scenarios of the DP shuttle tanker collision

Probability	Detection	Diagnosis	Decision-making and execution		
			Auto/Manual DC[a]	Recovery actions	Auto/Manual keep positioning
5.50×10^{-1}	Alarm detected	Failure in control situation			
1.09×10^{-1}	Alarm detected	Control situation performed perfectly	Failed auto DC[a]		
6.40×10^{-2}	Alarm detected	Control situation performed perfectly	Failed manual DC[a]		
1.24×10^{-4}	Alarm detected	Control situation performed perfectly	Manual DC[a]	Failure to start equipment on alternative switchboard	
1.24×10^{-4}	Alarm detected	Control situation performed perfectly	Manual DC[a]	Failure to start new generator	
2.85×10^{-5}	Alarm detected	Control situation performed perfectly	Manual DC[a]	Start of equipment on alternative switchboard	Failure of manual keep position
2.85×10^{-5}	Alarm detected	Control situation performed perfectly	Manual DC[a]	Start new generator	Failure of manual keep position
1.66×10^{-5}	Alarm detected	Control situation performed perfectly	Manual DC[a]	Start of equipment on alternative switchboard	Failure of automatic keep position
1.66×10^{-5}	Alarm detected	Control situation performed perfectly	Manual DC[a]	Start new generator	Failure of automatic keep position
1.25×10^{-6}	Alarm detected	Control situation performed perfectly	Manual DC[a]	Failure of tuning software	

[a] DC refers to disconnection

The scenario that occurred and is presented in the investigation report was the second most probable scenario out of 271 possible failure scenarios. In addition, according to the second and third most probable scenarios, automatic disconnection was more likely to fail than manual, mainly because the DP systems (thrusters, etc.) did not work properly. As a result, the failure probability of automatic modes may be deemed higher. However, after recovery actions, the DP systems work properly. Therefore, manual keep positions are more likely to fail than automatic mode, as presented in the table.

7.3 Application 3: Loss of Position of a DP Shuttle Tanker

On August 5, 2007, all of the position reference systems of a shuttle tanker were lost. Action was taken and the vessel was manually controlled by the DP operator. The vessel was kept steadily close to the ideal position/distance from a floating production storage and offloading (FPSO) unit. To safeguard the situation, the emergency shutdown (ESD) was activated. According to the investigation report, the vessel had not exceeded the maximum operation distance limit when ESD was activated.

The loading operation was stopped for 30 min while the DP system was taken to standby mode to allow the position reference system to be reset. Subsequently, all systems were found to be stable before the vessel was put back in DP mode.

When ESD was activated, the vessel communicated with the FPSO unit and informed that all position reference systems had been lost. The FPSO asked when the vessel was ready to resume cargo operations. Once the systems had been recalibrated, reset and deemed stable and the vessel was put back in DP mode, the master adjudged that loading could be resumed under the current conditions and under the close monitoring of the DP systems. After having informed the FPSO accordingly, and in agreement with the FPSO, loading operation was resumed [14].

This application is modeled based on the cognitive concept presented in Chap. 5, Subsect. 5.5. Figure 7.3 presents the possible events at each step of the dynamic event sequence diagram. The sequence of events, in the presented case study, is highlighted (gray boxes) in Fig. 7.3.

In this application, it is assumed that we only have the initial incident and ambient conditions, and we would like to generate more probable failure scenarios as the model output. In order to model, this incident based on the proposed DPRA methodology in Chap. 2, Subsect. 2.5.2; the following inputs are considered:

- Reference systems are lost.
- The operator has enough fitness for duty and training, and the stress level is normal.
- Environmental conditions including wind and waves (force and directions) are in normal conditions.
- Propulsion and computer systems work properly.

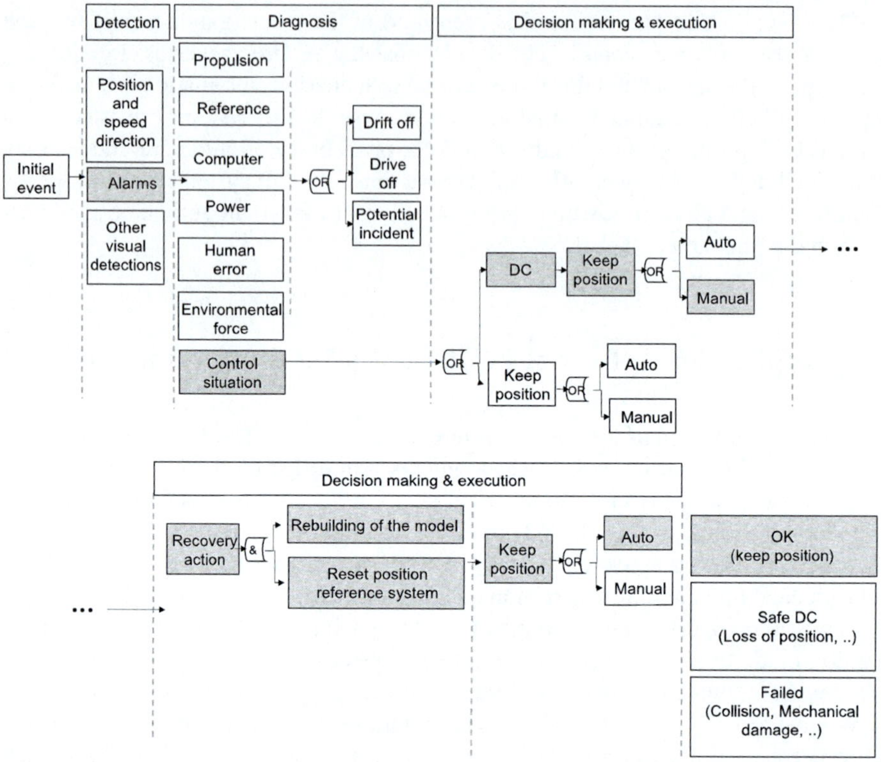

Fig. 7.3 Event sequence diagram of loss of position of the DP shuttle tanker, gray blocks represent the event sequence diagram of the incident, (DC: disconnection)

The most probable failure scenarios, considering the above-mentioned inputs, are calculated [7, 8, 11–13], are presented in Table 7.6, and the sum of all possible failure scenarios is equal to 0.422, meaning that the overall failure probability of the system is equal to 0.442.

According to the table, disconnection is performed in almost all of the most probable scenarios. This is mainly because of the fault of the reference systems. In such situations, based on expert judgment, it is more likely that the operator performs disconnection rather than trying to maintain position. Furthermore, the failure probability of automatic disconnection (first scenario) is higher in comparison with the manual mode (second scenario) due to the failed reference system. When the reference system has failed, accurate data cannot be transferred to the DP system. As a result, the computer and control system receive faulty data and the probability of failure increases.

In the manual mode, some of the faulty data would be modified by an operator's visual detections, and the probability of failure would be lower. After performing recovery action, however, DP systems including the reference system return to normal

Table 7.6 Most probable failure scenarios of loss of position of the dynamic positioning shuttle tanker

Probability	Detection	Diagnosis	Decision-making and execution		
			Auto/Manual DC[a]	Recovery actions	Auto/Manual keep positioning
2.4×10^{-1}	Alarm detected	Control situation performed perfectly	Failed automatic DC[a]		
1.42×10^{-1}	Alarm detected	Control situation performed perfectly	Failed manual DC[a]		
1.97×10^{-2}	Alarm detected	Failure in control situation			
6.32×10^{-3}	Alarm detected	Control situation performed perfectly	Manual DC[a]	Reference system recovery	Failure of manual keep position
6.32×10^{-3}	Alarm detected	Control situation performed perfectly	Manual DC[a]	Deselect faulty sensor	Failure of manual keep position
6.32×10^{-3}	Alarm detected	Control situation performed perfectly	Manual DC[a]	Recalibrate reference origin	Failure of manual keep position
6.32×10^{-3}	Alarm detected	Control situation performed perfectly	Manual DC[a]	Change position reference	Failure of manual keep position
3.19×10^{-3}	Alarm detected	Control situation performed perfectly	Manual DC[a]	Tuning software	Failure of manual keep position
1.91×10^{-3}	Alarm detected	Control situation performed perfectly	Manual DC[a]	Deselect faulty sensor	Failure of automatic keep position
1.91×10^{-3}	Alarm detected	Control situation performed perfectly	Manual DC[a]	Recalibrate reference origin	Failure of automatic keep position

[a] DC refers to disconnection

operation. Therefore, the failure probability of manually keeping position is higher than the automatic mode.

7.4 Application 4: Collision Between a Supply Ship and a Petroleum Installation

On June 7, 2019, a collision occurred between a supply ship and a petroleum installation during loading/discharging. In this accident, a technical fault caused the vessel's load-reduction mode to be activated, reducing the power to all of its thrusters to 10–15 percent of the maximum. Later, power was lost for two out of three bow thrusters. The vessel thereby lost position. The officer attempted to switch the vessel to partly manual positioning. The vessel drifted against the facility, suffering extensive damage to the mast and equipment above the bridge, and denting its starboard side aft. Moreover, the mast caused damage to the installation's lifeboat station. The master switched from DP to manual positioning. The programmable logic controller (PLC) was then reset to blackout safety system generators 2 and 4.

The direct causes to the drift off incident were inadequate thruster power, and the location of the loading/discharge operation on the windward (weather) side. The underlying causes that lead to insufficient thruster power were related to the failure of the incorrect installation of the equipment components, with disruption from the defective components leading to network failure in the blackout safety system ("network storm"), a loss of network frequency measurement on the main switchboard, activation of the load-reduction mode, and restriction of all thrusters to 10–15 percent of maximum output, nonconformity between DP commands, and revolutions per minute (RPM) feedback from all thrusters, and automatic shutdown of thrusters 1 and 3 [15]. This application is modeled based on the cognitive concept presented in Chap. 5, Subsect. 5.5. The sequence of events, in the presented case study, are highlighted (gray boxes) in Fig. 7.4.

According to the investigation report, the initial event and conditions of this accident were as follows:

- The operator had enough fitness for duty and training, and the stress level was normal.
- There was not enough time, and the vessel was in an emergency situation.
- Power was reduced to 15%.
- The reference and computer systems worked properly.
- The propulsion system including thrusters was not working properly.

These conditions are utilized here as model inputs, and failure scenarios are generated as results [7, 8, 11–13]. Table 7.7 presents the most probable failure scenarios of the supply ship.

As can be seen, the scenario that occurred and is presented in the investigation report is the third most probable scenario among 271 possibilities. In addition,

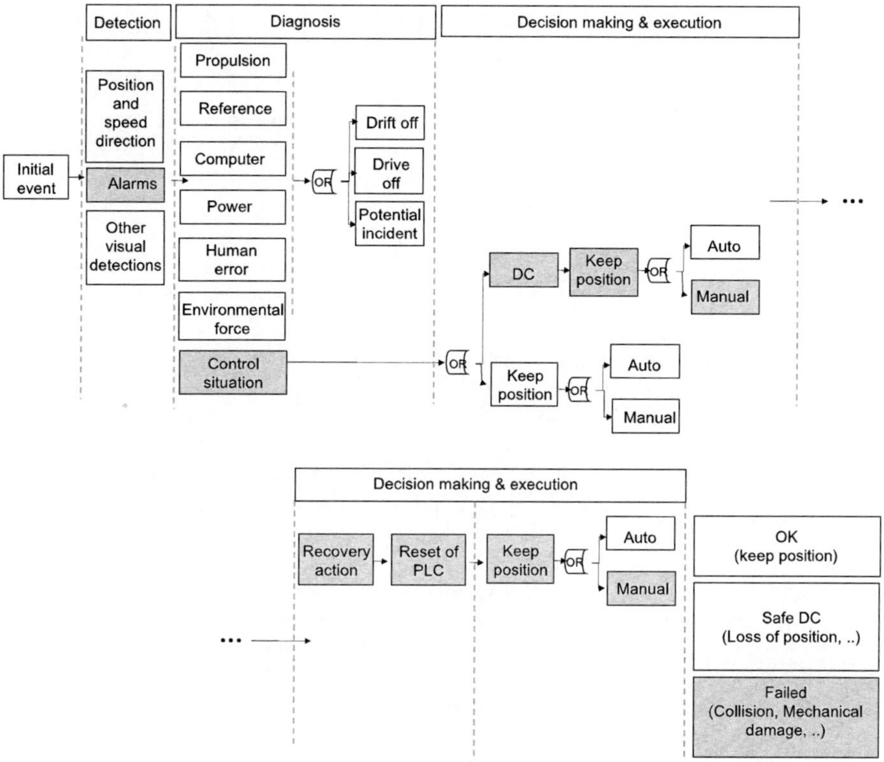

Fig. 7.4 Event sequence diagram of the supply ship collision, gray blocks represent the event sequence diagram of the accident, (DC: disconnection, PLC: programmable logic controller)

according to the second and third most probable scenarios, automatic disconnection is more likely to fail than manual. This matter can be inferred from the rest of the scenarios, as most of the successful disconnections are performed manually. This is mainly because DP systems (power system and thrusters) do not work properly in this incident. As a result, the failure probability of the automatic mode is higher. However, after performing recovery actions, the DP systems work properly. Therefore, manually keeping position is more likely to fail than the automatic mode, as presented in the table.

In these four applications, different DP accidents were modeled using methods presented in the previous chapters. The accidents' characteristics and causes have been explored using the information from the investigation reports of each accident, and the model outputs have been compared with the accident scenarios presented in the investigation reports. According to these comparisons, the proposed methods are able to predict the most probable failure scenarios with a high degree of accuracy, which is an important part of online risk assessment.

Table 7.7 Most probable failure scenarios of the supply ship collision

Probability	Detection	Diagnosis	Decision-making and execution		
			Auto/Manual DC[a]	Recovery actions	Auto/Manual keep positioning
5.53×10^{-1}	Alarm detected	Failure in control situation			
1.07×10^{-1}	Alarm detected	Control situation performed perfectly	Failed automatic DC[a]		
6.38×10^{-2}	Alarm detected	Control situation performed perfectly	Failed manual DC[a]		
1.23×10^{-4}	Alarm detected	Control situation performed perfectly	Manual DC[a]	Failure to start new generator	
1.23×10^{-4}	Alarm detected	Control situation performed perfectly	Manual DC[a]	Failure to start equipment on alternative switchboard	
2.84×10^{-5}	Alarm detected	Control situation performed perfectly	Manual DC[a]	Start new generator	Failure of manual keep position
2.84×10^{-5}	Alarm detected	Control situation performed perfectly	Manual DC[a]	Start of equipment on alternative switchboard	Failure of manual keep position
1.42×10^{-5}	Alarm detected	Control situation performed perfectly	Manual DC[a]	Start new generator	Failure of automatic keep position
1.42×10^{-5}	Alarm detected	Control situation performed perfectly	Manual DC[a]	Start of equipment on alternative switchboard	Failure of automatic keep position
6.41×10^{-6}	Alarm detected	Control situation performed perfectly	Automatic DC[a]	Failure to start equipment on alternative switchboard	

[a] DC refers to disconnection

References

1. Hakonsund K (2018) Investigation report level 2, Rig excursion reached red watch circle
2. Bye A, Laumann K, Taylor C, Rasmussen M, Øie S, van de Merwe K, Gould K (2017) The petro-HRA guideline
3. Cruz DF, Fonseca DDR (2017, Oct) WSOG and emergency disconnection guidelines. In: OTC Brazil. Offshore technology conference
4. Chen H, Moan T, Verhoeven H (2008) Safety of dynamic positioning operations on mobile offshore drilling units. Reliab Eng Syst Saf 93(7):1072–1090
5. Hogenboom S, Parhizkar T, Vinnem JE (2021) Temporal decision-making factors in risk analyses of dynamic positioning operations. Reliab Eng Syst Saf 207:107347
6. Parhizkar T, Utne IB, Vinnem JE, Mosleh A (2021) Dynamic probabilistic risk assessment of decision-making in emergencies for complex systems, case study: Dynamic positioning drilling unit. Ocean Eng 237:109653
7. Parhizkar T, Vinnem JE, Utne IB, Mosleh A (2020) Supervised dynamic probabilistic risk assessment of complex systems, Part 1: general overview. Reliab Eng Syst Saf 107406
8. Parhizkar T, Utne IB, Vinnem JE, Mosleh A (2021) Supervised dynamic probabilistic risk assessment of complex systems, Part 2: application to risk-informed decision making, practice and results. Reliab Eng Syst Saf 208:107392
9. Parhizkar T, Hogenboom S, Vinnem JE, Utne IB (2020) Data driven approach to risk management and decision support for dynamic positioning systems. Reliab Eng Syst Saf 201:106964
10. Members/NBN T (2006) Significant incident investigation XXX DP incident on XXX. confidential report. Teekay shipping (Canada) Ltd
11. Hu Y, Parhizkar T, Mosleh A (2021) Guided simulation for dynamic probabilistic risk assessment of complex systems: concept, method, and application. Reliab Eng Syst Saf 108047.
12. Parhizkar T, Mosleh A (2022) Guided probabilistic simulation of complex systems toward rare and extreme events. The 68th Annual Reliability & Maintainability Symposium (RAMS), Tucson, Arizona
13. Hamed SN, Parhizkar T, Mosleh A (2021) Simulation based probabilistic risk assessment (simpra): risk based design. 31st European Safety and Reliability Conference, 19-23 September 2021, Angers, France
14. Lund T (2007) Confidential incident investigation, loss of DP at XXX. TeekayMarine
15. Leonhardsen R (2019) Investigation of collision between Sjoborg supply ship and Statfjord A, petroleum safety authority norway

Chapter 8
Challenges of Online Dynamic Probabilistic Risk Assessment and Possible Solutions

8.1 Risk Assessment Challenges with Autonomous Systems

Autonomous systems and technologies should contribute to safer and more efficient operations. Software and advanced control systems in these systems with learning capabilities, however, also involve complexities that challenge the identification of hazardous events, assessment, and mitigation of risks [1, 2].

Societal acceptance of autonomous systems depends on public trust in the technology and in the regulators that give approval. Developers and producers must be able to ensure the safe, reliable and secure performance of autonomous systems and functions and compliance with requirements. Operators of the systems need to be able to execute safe and efficient operations [3–9], with efficient risk control. Authorities need standards and guidelines for autonomy as a basis for their regulatory and inspection activities. Acceptance and approval of autonomous systems require efficient risk identification, assessment, modelling, testing, verification, and validation in the design process, and during operation [1].

Autonomous systems are cyber physical systems and may use Artificial Intelligence (AI) as part of their decision making [1]. For systems with a low level of autonomy (LoA), situation awareness of the surroundings and system integrity are mainly related to alarm systems and the human operator's perception and understanding of the system and operation. To improve the intelligence and achieve higher LoA, the system itself needs to be able to plan and re-plan the mission which means that situation awareness (SA) needs to be "transferred" from the human operator to the autonomous system itself, including learning abilities and decision making. This means that to design and utilize systems that cooperate, possibly replace, and outperform human capabilities, means that supervisory risk control is decisive to achieve an acceptable risk level, [10].

Supervisory risk control is a dynamic functionality that should be designed and implemented into the autonomous system's control system to enable the system to assess and control risks during operation. A critical part of supervisory risk control is

T. Parhizkar et al., *Online Probabilistic Risk Assessment of Complex Marine Systems*, Springer Series in Reliability Engineering,
https://doi.org/10.1007/978-3-030-88098-9_8

development of online risk assessment and modelling [10], which encounters challenges related to the availability of effective and adequate risk assessment methods. Thieme et al. [11] analyzed 64 collision risk models for maritime autonomous surface ships (MASS) and found that none of them included analysis of the communication connection with a shore base, which is important for autonomous ships which have reduced or no crew onboard. Interaction between conventional and autonomous ships, as well as potential changes in traffic patterns, are other aspects that needs to be taken into consideration in the risk assessments.

There are challenges with integrated risk modelling of hardware, human, and software failures, analysis of learning systems, and available data [1, 12–14]. This book, however, is an attempt to aid in the development of online risk assessment, not only for providing decision support for human operators and supervisors of autonomous systems, but also for supervisory risk control. Even though methods and ways for handling some of the challenges are suggested, further work is needed in several areas, of which some are discussed in the following sub-chapters.

8.2 Online Data Collection

Online models are connected to a database, and data are updated frequently, according to the system boundary and goals [15, 16]. The process of collecting, storing and updating data should be aligned with the connected analysis model. In dynamic probabilistic risk assessment (DPRA) models, the risk level of the system should be updated in a short time span, e.g., from seconds to weeks, depending on the operation and systems involved, [17, 18]. The time span is dependent on the type of industry for which the model is applied, and the specific accident scenario studied. Consequently, the time interval of updating data should be specified accordingly. Available cloud servers and engines to store and update data are fast enough to meet the requirements of DPRA [19, 20]. However, the main challenge is the data collection phase.

As mentioned in previous chapters, a comprehensive DPRA model considers environmental, human, organizational, software, and hardware factors´ interactions in the risk assessment process [21–24]. To meet this criterion, we should be able to measure the main characteristics of human, organization, and technical components [25, 26]. Some of the characteristics are more or less static (depending on the time horizon) and their values are quantified once at the beginning of the risk assessment and modeling. These parameters remain constant over time. But there are other more dynamic characteristics, and their values change over time. These characteristics shape the dynamic behavior of a complex system. Dynamic parameters are quantified using online measurements over time [27, 28]. At this step of modeling, some simplification should be performed as all system dynamic characteristics could not be quantified straightforwardly. For instance, human performance shaping factors (PSFs) such as stress level cannot be quantified easily, and some assumptions need to be applied. Other PSFs, including training and experience level, could be measured and updated depending on the shift´s operators [29, 30].

Online measurements of dynamic parameters of human, organizational, and technical factors should be quantified for DPRA models. There are different methods of measurement and quantification, depending on the parameter of the study, goal of the model and available budget. For instance, operating conditions of technical components could be quantified using sensors with different accuracy and costs [31, 32]. The organizational and human factors could be quantified using, observations, interviews, and questionnaires [33–35].

A general framework should be developed to specify different types of measurements of dynamic parameters in probabilistic risk assessment, considering different system goals and available budgets. This framework should provide required static and dynamic parameters for probabilistic risk assessment of a complex system. The designer should simplify and specify the measurement methods for the complex system under study, considering system functionality, restrictions, required model accuracy, and available budget.

8.3 Uncertainties

Probabilistically risk assessment of complex systems could be of infinite size depending on the nature of the physical phenomena involved and the resolution needed. Risk assessment models are subject to stochastic or aleatory uncertainty, and knowledge-driven or epistemic uncertainty [36, 37].

Aleatory uncertainty is the natural randomness in a process. It refers to the inherent uncertainty due to the probabilistic variability. This uncertainty is characterized by a probability distribution. For discrete variables, the randomness is parameterized by the probability of each possible value. For continuous variables, the randomness is parameterized by the probability density function [38, 39].

Epistemic uncertainty is uncertainty due to limited data and knowledge. This lack of knowledge comes from many sources. Inadequate understanding of the underlying processes, incomplete knowledge of the phenomena, or imprecise evaluation of the related characteristics are common sources of epistemic uncertainty. The epistemic uncertainty may be characterized by alternative models. For discrete random variables, the epistemic uncertainty can be modelled by alternative probability distributions. For continuous random variables, the epistemic uncertainty can be modelled by alternative probability density functions. In addition, there is epistemic uncertainty in parameters resulting from unknown values in the system [38, 39].

In DPRA models, in addition to model uncertainties, the aleatory and epistemic uncertainties exist in the input parameters (e.g., components failure frequencies) of the model. For instance, in the presented examples in the book related to the dynamic positioning (DP) system, in Chaps. 3 and 7, the study and parts of the analyses heavily rely on the contents of the international marine contractor's association (IMCA) reports and the information available. However, it is difficult to verify the quality of the incident reports, since incidents are not in their original form. Other reasons could be that the original incident investigation was not detailed enough to clarify causal factors or that it was not conducted by people with the right competencies.

Another common limitation related to the data quality is that the description of incidents in the IMCA reports is lacking in detail, leading to interpretation and subjectivity.

Moreover, underreporting has to be assumed. The IMCA reports are based on incidents that have been shared by various vessel owners voluntarily. It can often be assumed that they are intentionally not comprehensive representation of the incident.

How much of the incidents are missing from the IMCA reports can only be guessed. Since the incident reports for the case studies are used to exemplify the models and methods, the data quality plays a different role. The incident reports are in their original form and provided by the vessel's operator. However, the level of detail needed, for example, for a proper human reliability analysis is insufficient in these reports.

Furthermore, there are non-observable data, such as behavioral or mental states (belief, desire, etc.) of an operator that affect human error. These types of data may impact the risk level of the system. In the presented case studies, a Bayesian network (BN) is used for inferring latent variables such as belief and desire. The parent nodes of BN that take input parameters, such as stress level or fitness for duty of the operator, could include data uncertainty by using distributions for the states.

To identify the main sources of uncertainties and attempt to quantify them is one of the first steps of DPRA models. In recent years, multiple algorithms have been developed to quantify uncertainties of risk assessments [40, 41]. However, identifying the source and nature of uncertainties of a complex system could be a very complicated process, that should be performed by analysts. Developing a general procedure to define the main uncertainties in a complex system when performing DPRA could assist analysts significantly.

However, in this regard, it should be noted that monitoring the risk level under uncertainty and capturing sudden changes in system behavior could assist operators and the control system in responding to incidents. In other words, understanding the relative risk level could be more informative than the absolute certain risk value itself. Developing online risk models under all available uncertainties still help operators to detect and diagnose the incident. In addition, input to decision support tools is often a comparison of two or more alternatives. Uncertainties in a relative comparison is lower for absolute numbers.

Nevertheless, understanding and analyzing the uncertainty sources will improve the online risk models' accuracy and provide operators with a better picture of system risk level over time, and hence improve the decision support.

8.4 Complexity and Execution Time

As presented in [42], a major challenge is related to the use of the DPRA for a complex system. Dynamic risk assessment is well stablished for cases with considerable data and system behavior information [43, 44]. In this regard, multiple statistical and probabilistic tools can be used to provide valuable assessment of dynamic probabilistic risk level in different applications [45–47]. However, DPRA is primarily regarding situations characterized by large uncertainties. These situations necessitate the assessment of multiple alternative possible scenarios and, occasionally, unknown scenarios [48].

In multiple studies, it is shown that the number of possible alternative scenarios increase over time significantly [48, 49]. This matter will increase the execution time dramatically, and in some cases make the problem unsolvable.

In Chap. 2 (Sect. 2.5), different methods to control the size of the DPRA scenarios over time is proposed. One of the presented methods is a supervised DPRA model that generates and sorts failure scenarios in a short execution time using optimization models. Hence, this method is highly valuable for risk-informed decision making in emergencies and situations with time restrictions. In Chap. 7, the application of this supervised DPRA model has been examined and discussed. In this model, knowledge of the system is explicitly used to predict possible failure scenarios. Further, optimization in terms of a supervised learning algorithm is used to find the optimal solution, ranking of the most probable failure scenarios. This means that instead of focusing on obtaining all possible scenarios, we have approached the problem of exploring the most likely failure scenarios efficiently. The model has been applied to solve DPRA problems in four different DP incidents with promising results. A potential challenge with this approach is that low probability—high consequence scenarios may be concealed. Further work on the generation and selection of scenarios is ongoing, as presented in [50].

Risk-informed decision-making has to deal with, to an increasing extent, unknown situations, such as incidents in a self-learning complex system. In self-learning systems, the behavior of the system and its responses to the environment is updated/changed over time, e.g., the system response over time to a specific incident will not be the same as its response in the beginning of its operation [51, 52].

In these cases, system behavior and principal rules are updated continuously. Therefore, the consequences and recovery plans of an incident will change over time [53]. The available probabilistic dynamic risk approaches and methods are not capturing self-learning behavior of complex systems. Although risk assessment methods use predictive methods to evaluate the risk level of the system, the methods need to be extended to assist in predicting self-learning system behavior over time. In this regard, predictive methods and analysis in self-learning systems can be considered as new approaches for improving DPRA methods.

In addition, according to the analysis results presented in Chap. 7, available time has a greatest effect on the risk level of a complex system operation in emergency situations. Therefore, managing the available time may significantly reduce the risk level of a system. This is important for autonomous systems, both with reduced manning and with human supervisors in a shore control center. Fault diagnosis methods such as artificial intelligence (AI) may provide useful information to operators by analyzing and interpreting available data [54, 55]. With the development of AI, many new methods have been introduced to research into the fault diagnosis of complex systems [56, 57]. Utilizing these methods may enable operators to determine the root cause of failure, and consequently to significantly reduce the time required for the diagnosis phase.

Furthermore, prediction models could help operators to make a better decision within a limited timeframe [58–60], e.g., in emergency situations, there are multiple alternative decision scenarios that can be made to control the hazard. Predictive simulation models could be employed to simulate each decision scenario and achieve a better overview of their consequences. Fault diagnosis methods and prediction models for risk assessment of complex systems should be explored in future works to help operators make better decisions in emergency situations.

8.5 Applicability and Connection to Decision Support

One of the main paths for future research related to online risk assessment is the potential in advancing and coupling the DPRA methodology to online decision support tools. Such tools support analysts and decision makers in making better and/or faster decisions [61, 62]. For autonomous systems, online risk assessment may also support decision making by the control system, and as such, enable higher autonomy, which is the focus and aim of supervisory risk control.

Online decision support may provide a better risk picture and allow for decision making, moving from a re-active emergency management to a pro-active approach where decision scenarios could be analyzed and anticipated based on real-time information. One such possibility would be to simulate all alternative decision scenarios and provide an approximate time scale and safety level of each scenario. Another option would be to allow the operator(s) to be informed about the event and sequence probabilities in the near future. Knowing this information could help operator(s) to readjust system configuration with improved knowledge about the system.

The usefulness of coupling DPRA methodology and decision support may be seen in the framework presented in Chap. 3. The main decision that a dynamic positioning operator (DPO) has to make is whether or not she/he can safely remain on location or needs to move off. In this case, the DPRA model could provide failure probabilities and risk levels of these two alternative scenarios that could help operators to make safer decisions.

Other potential uses could include maintenance planning. When planning maintenance of the main components of the DP system, possible risk effects of delayed or postponed maintenance tasks could be analyzed with a DPRA methodology.

Last but not least, in training of DP operators could DPRA methodology be used to add insights into effects of operator actions and potential consequences. DPRA methodology could be used to illustrate the risk level implied by various actions, and the effects of response time variations could be illustrated. Some of these aspects are discussed by Hogenboom et. al. [63].

Even though this book focuses on complex marine systems, with particular focus on the DP system, DPRA and online risk assessment and modeling are valuable for providing decision support in other types of automated and autonomous systems, as well.

References

1. Thieme C, Ramos MA, Utne IB, Mosleh A (2021) Proceedings of the international workshop on autonomous systems safety. Online event, https://www.ntnu.edu/imt/iwass
2. Thieme C, Ramos MA, Utne IB, Mosleh A (2021) Proceedings of the international workshop on autonomous systems safety, online event
3. Parhizkar T, Roshandel R (2017) Long term performance degradation analysis and optimization of anode supported solid oxide fuel cell stacks. Energy Convers Manage 133:20-30
4. Roshandel R, Parhizkar T (2016) Degradation based optimization framework for long term applications of energy systems, case study: solid oxide fuel cell stacks. Energy 107:172-181
5. Parhizkar T, Hafeznezami S (2018) Degradation based operational optimization model to improve the productivity of energy systems, case study: Solid oxide fuel cell stacks. Energy Convers Manage 158:81-91
6. Roshandel R, Parhizgar T (2013) A new approach to optimize the operating conditions of a polymer electrolyte membrane fuel cell based on degradation mechanisms. Energy Syst 4(3):219-237
7. Parhizkar T (2018) Long-term degradation-based modeling and optimization framework. In: Handbook of research on predictive modeling and optimization methods in science and engineering (pp 192-220). IGI Global
8. Parhizkar T, Mosleh A, Roshandel R (2017) Aging based optimal scheduling framework for power plants using equivalent operating hour approach. Applied energy 205:1345-1363
9. Sotoodeh AF, Parhizkar T, Mehrgoo M, Ghazi M, Amidpour M (2019) Aging based design and operation optimization of organic rankine cycle systems. Energy Convers Manage 199:111892
10. Utne IB, Rokseth B, Sørensen AJ, Vinnem JE (2020) Towards supervisory risk control of autonomous ships. Reliab Eng Syst Saf 196:106757
11. Thieme CA, Utne IB, Haugen S (2018) Assessing ship risk model applicability to marine autonomous surface ships. Ocean Eng 165:140–154
12. Hu Y, Parhizkar T, Mosleh A (2021) Guided simulation for dynamic probabilistic risk assessment of complex systems: concept, method, and application. Reliab Eng Syst Saf 108047
13. Parhizkar T, Mosleh A (2022) Guided probabilistic simulation of complex systems toward rare and extreme events. The 68th Annual reliability & maintainability symposium (RAMS), Tucson, Arizona
14. Hamed SN, Parhizkar T, Mosleh A (2021) Simulation based probabilistic risk assessment (simpra): risk based design. 31st European Safety and Reliability Conference, 19-23 September 2021, Angers, France
15. Martyn M (2003) The hybrid online model: good practice. Educ Q 26(1):18–23
16. Parhizkar T, Hogenboom S, Vinnem JE, Utne IB (2020) Data driven approach to risk management and decision support for dynamic positioning systems. Reliab Eng Syst Saf 201:106964
17. Årnes A, Sallhammar K, Haslum K, Brekne T, Moe MEG, Knapskog SJ (2005) Real-time risk assessment with network sensors and intrusion detection systems. In: International conference on computational and information science. Springer, Berlin, Heidelberg, pp 388–397

18. Meel A, Seider WD (2008) Real-time risk analysis of safety systems. Comput Chem Eng 32(4–5):827–840
19. Lindström J, Raatikka V, Ruuth J, Soini P, Vakkila K (2013) IBM solidDB: in-memory database optimized for extreme speed and availability. IEEE Data Eng Bull 36(2):14–20
20. Tremaine RB, Franaszek PA, Robinson JT, Schulz CO, Smith TB, Wazlowski ME, Bland PM (2001) IBM memory expansion technology (MXT). IBM J Res Dev 45(2):271–285
21. Skogdalen JE, Vinnem JE (2011) Quantitative risk analysis offshore-human and organizational factors. Reliab Eng Syst Saf 96(4):468–479
22. Aven T, Sklet S, Vinnem JE (2006) Barrier and operational risk analysis of hydrocarbon releases (BORA-Release): part I. Method description. J Hazard Mater 137(2):681–691
23. Thieme CA, Mosleh A, Utne IB, Hegde J (2020) Incorporating software failure in risk analysis—part 1: software functional failure mode classification. Reliab Eng Syst Saf 197:106803
24. Thieme CA, Mosleh A, Utne IB, Hegde J (2020) Incorporating software failure in risk analysis—part 2: risk modeling process and case study. Reliab Eng Syst Saf 198:106804
25. Parhizkar T, Utne IB, Vinnem JE, Mosleh A (2021) Dynamic probabilistic risk assessment of decision-making in emergencies for complex systems, case study: Dynamic positioning drilling unit. Ocean Eng 237:109653
26. Parhizkar T, Utne IB, Vinnem JE, Mosleh A (2021) Supervised dynamic probabilistic risk assessment of complex systems, Part 2: application to risk-informed decision making, practice and results. Reliab Eng Syst Saf 208:107392
27. Chang YHJ, Mosleh A (2007) Cognitive modeling and dynamic probabilistic simulation of operating crew response to complex system accidents. Part 2: IDAC performance influencing factors model. Reliab Eng Syst Saf 92(8):1014–1040
28. Laumann K (2020) Criteria for qualitative methods in human reliability analysis. Reliab Eng Syst Saf 194:106198
29. Ergan S, Radwan A, Zou Z, Tseng HA, Han X (2019) Quantifying human experience in architectural spaces with integrated virtual reality and body sensor networks. J Comput Civ Eng 33(2):04018062
30. Al-Shargie F (2019) Quantification of mental stress using fNIRS signals
31. Parhizkar T, Balali S, Mosleh A (2018) An entropy based bayesian network framework for system health monitoring. Entropy 20(6):416
32. Bein D, Jolly V, Kumar B, Latifi S (2005) Reliability modeling in wireless sensor networks. Int J Inf Technol 11(2):1–8
33. Stanton N, Salmon PM, Rafferty LA (2013) Human factors methods: a practical guide for engineering and design. Ashgate Publishing, Ltd.
34. Xenikou A, Furnham A (1996) A correlational and factor analytic study of four questionnaire measures of organizational culture. Human Relations 49(3):349–371
35. Putz D, Schilling J, Kluge A, Stangenberg C (2013) Measuring organizational learning from errors: development and validation of an integrated model and questionnaire. Manag Learn 44(5):511–536
36. Hora SC (1996) Aleatory and epistemic uncertainty in probability elicitation with an example from hazardous waste management. Reliab Eng Syst Saf 54(2–3):217–223
37. Dolšek M (2012) Simplified method for seismic risk assessment of buildings with consideration of aleatory and epistemic uncertainty. Struct Infrastruct Eng 8(10):939–953
38. Rao KD, Kushwaha HS, Verma AK, Srividya A (2007) Quantification of epistemic and aleatory uncertainties in level-1 probabilistic safety assessment studies. Reliab Eng Syst Saf 92(7):947–956
39. Bae HR, Grandhi RV, Canfield RA (2004) An approximation approach for uncertainty quantification using evidence theory. Reliab Eng Syst Saf 86(3):215–225
40. Abdo H, Flaus JM, Masse F (2017) Uncertainty quantification in risk assessment-representation, propagation and treatment approaches: application to atmospheric dispersion modeling. J Loss Prev Process Ind 49:551–571

41. Abdo H, Flaus JM (2016) Uncertainty quantification in dynamic system risk assessment: a new approach with randomness and fuzzy theory. Int J Prod Res 54(19):5862–5885
42. Mosleh A (2014) PRA: a perspective on strengths, current limitations, and possible improvements. Nucl Eng Technol 46(1):1–10
43. Khakzad N, Khan F, Amyotte P (2012) Dynamic risk analysis using bow-tie approach. Reliab Eng Syst Saf 104:36–44
44. Kalantarnia M, Khan F, Hawboldt K (2009) Dynamic risk assessment using failure assessment and Bayesian theory. J Loss Prev Process Ind 22(5):600–606
45. Parhizkar T, Vinnem JE, Utne IB, Mosleh A (2020) Supervised dynamic probabilistic risk assessment of complex systems, Part 1: general overview. Reliab Eng Syst Saf 107406
46. Zhao J, Jin J, Zhang X, Chen Y (2012) Dynamic risk assessment model for water quality on projection pursuit cluster. Hydrol Res 43(6):798–807
47. Villa V, Paltrinieri N, Khan F, Cozzani V (2016) Towards dynamic risk analysis: a review of the risk assessment approach and its limitations in the chemical process industry. Saf Sci 89:77–93
48. Coyne K (2009) A predictive model of nuclear power plant crew decision-making and performance in a dynamic simulation environment (Doctoral dissertation). University of Maryland, Mechanical Engineering Department
49. Diaconeasa MA (2017) Integration of qualitative and quantitative hybrid causal logic into a simulation-based platform for probabilistic risk assessment of nuclear power plants. PhD thesis, University of California, Los Angeles
50. Maidana RG, Parhizkar T, Thieme ChA, Ramos M, Utne I, Mosleh A (2021) Towards Risk-based autonomous decision-making with accident dynamic simulation, ESREL conference, France
51. Li C, Wu T, Chan FT (2012) Self-learning complex neuro-fuzzy system with complex fuzzy sets and its application to adaptive image noise canceling. Neurocomputing 94:121–139
52. Nguyen DH, Widrow B (1990) Neural networks for self-learning control systems. IEEE Control Syst Mag 10(3):18–23
53. Chatterjee K, Zufferey D, Nowak MA (2012) Evolutionary game dynamics in populations with different learners. J Theor Biol 301:161–173
54. Korbicz J, Koscielny JM, Kowalczuk Z, Cholewa W (eds) (2012) Fault diagnosis: models, artificial intelligence, applications. Springer Science & Business Media
55. Simani S, Fantuzzi C, Patton RJ (2003) Model-based fault diagnosis techniques. In: Model-based fault diagnosis in dynamic systems using identification techniques. Springer, London, pp 19–60
56. Frank PM, Köppen-Seliger B (1997) New developments using AI in fault diagnosis. Eng Appl Artif Intell 10(1):3–14
57. Fenton WG, McGinnity TM, Maguire LP (2001) Fault diagnosis of electronic systems using intelligent techniques: a review. IEEE Trans Syst Man Cybern Part C (Applications and Reviews) 31(3):269–281
58. Mosavi A (2015) Predictive decision making. Report, Technische Universität Darmstadt, Predictive Decision Model
59. Milovic B, Milovic M (2012) Prediction and decision making in health care using data mining. Kuwait Chap Arabian J Bus Manage Rev 1(12):126
60. Stangl DK (1995) Prediction and decision-making using Bayesian hierarchical models. Stat Med 14(20):2173–2190
61. Parhizkar T, Aramoun F, Saboohi Y (2020) Efficient health monitoring of buildings using failure modes and effects analysis case study: air handling unit system. J Build Eng 29:101113
62. Parhizkar T, Aramoun F, Esbati S, Saboohi Y (2019) Efficient performance monitoring of building central heating system using Bayesian Network method. J Build Eng 26:100835
63. Hogenboom S (2020) Decision-making for dynamic positioning operations in the offshore oil and gas industry—a human factors based approach. PhD thesis, NTNU, Trondheim, Norway

Chapter 9
Attachment A—Fault Trees for Different Dynamic Positioning Systems

In this attachment, the specific fault trees for different types of dynamic positioning (DP) systems are presented and compared.

DP platform supply vessels range from 50 to 100 m in length and are capable of carrying out a diversity of tasks. They primarily support offshore oil and gas platforms by transporting necessary supplies to and a platform from the base ashore. The failure modes of DP platform supply vessels are different from DP drilling units, which mostly are operating in deep-water drilling operations. It is vital for the drilling unit to stay on location over the oil/gas well, such that the riser connecting to the well is nearly upright and vertical. The lower main riser angle is continuously monitored and maintained up to accurate levels in order to avoid unwanted disconnections.

Presently, the DP rigs and vessels are configured in a way to operate in water depths of up to 3000 m or more with the help of differential global positioning system (DGPS) and long baseline (LBL) acoustic systems within the DP module. Figures 9.1, 9.2, 9.3, 9.4, 9.5 and 9.6 present propulsion, power, and reference system fault trees of DP platform supply vessel, respectively. Highlighted events (gray boxes) have the highest failure frequency among the rest of events in each fault tree. The computer system fault tree for supply vessels is the same as the general form of fault tree applicable to all types of DP systems.

The specific fault trees for the other DP types including drilling vessel, diving vessel, cable laying vessel, and shuttle tanker are presented in the following figures (Figs. 9.4, 9.5, 9.6, 9.7, 9.8, 9.9, 9.10, 9.11, 9.12, 9.13 and 9.14). It should be noted that the component fault trees presented for each type of DP systems are the ones that are different from the general form of the component fault trees (Figs. 2.4, 2.5, 2.6, 2.7, 2.8, 2.9 and 2.10). Figure 9.4, 9.5 and 9.6 present power, propulsion, and reference system fault trees of DP drilling unit, respectively.

Many DP vessels are specially designed for supporting commercial subsea operations such as diving. Diving support vessels are thus used for this purpose. Divers are required to carry out inspections or survey jobs, installations, and configurations of subsea equipment, monitoring an operation, and recovery of lost or abandoned

T. Parhizkar et al., *Online Probabilistic Risk Assessment of Complex Marine Systems*, Springer Series in Reliability Engineering,
https://doi.org/10.1007/978-3-030-88098-9_9

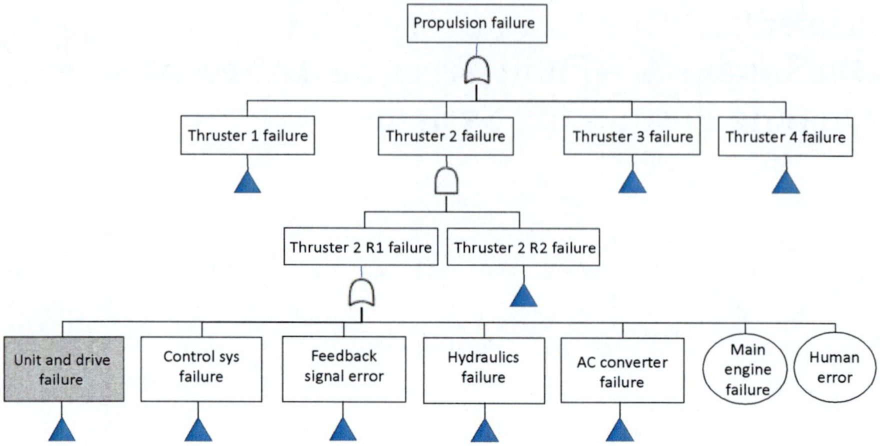

Fig. 9.1 Fault tree of a propulsion system in a dynamic positioning (DP) supply vessel, highlighted event has the highest failure frequency. (R1: redundant 1, R2: redundant 2, AC: alternating current)

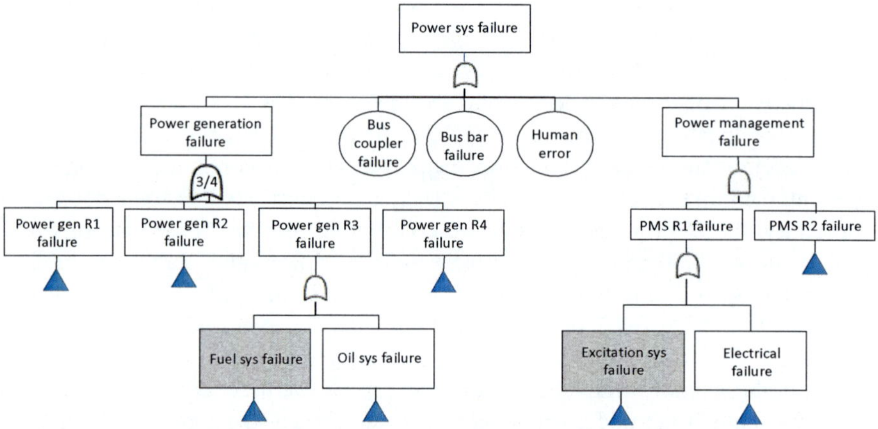

Fig. 9.2 Fault tree of a power system in a DP supply vessel, highlighted events have the highest failure frequencies. (Ri: redundant i, gen: generator, PMS: power management system, sys: system)

equipment. Figures 9.7 and 9.8 present propulsion and reference system fault trees of DP diving support vessels, respectively.

Remotely operated vehicles (ROVs) are gradually substituting the divers in many jobs, but there are still certain tasks that cannot be carried out remotely. Figure 9.9 presents the reference system fault tree of ROV support vessels.

Fiber-optic cables are used to connect the world through the "World Wide Web." To avoid heavy losses structurally to these cables, it is now common to use DP vessels for the cable laying and repairing jobs. Most of the current cable laying ships are installed with DP systems. Furthermore, many pipe lay operations are carried out

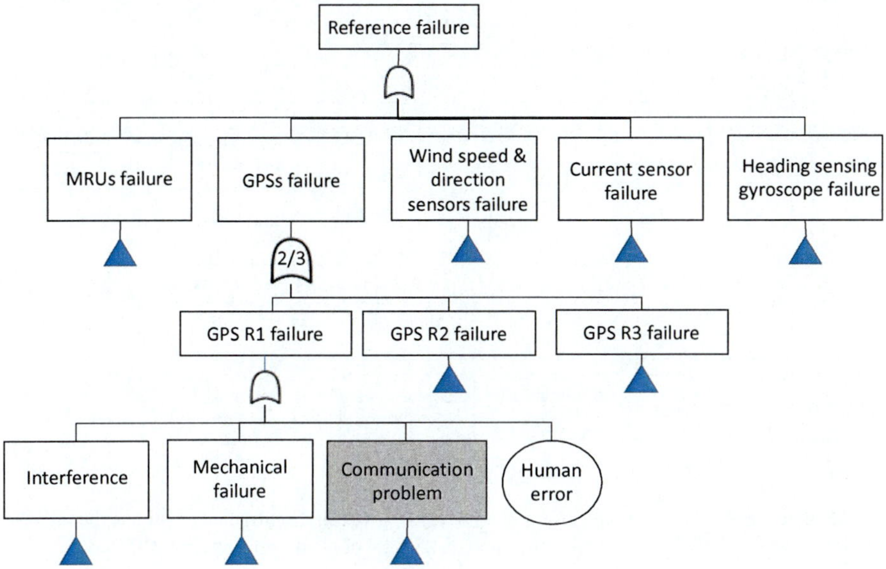

Fig. 9.3 Fault tree of a reference system in a DP supply vessel, highlighted event has the highest failure frequency. (MRU: motion reference unit, GPS: global positioning system, Ri: Redundant i)

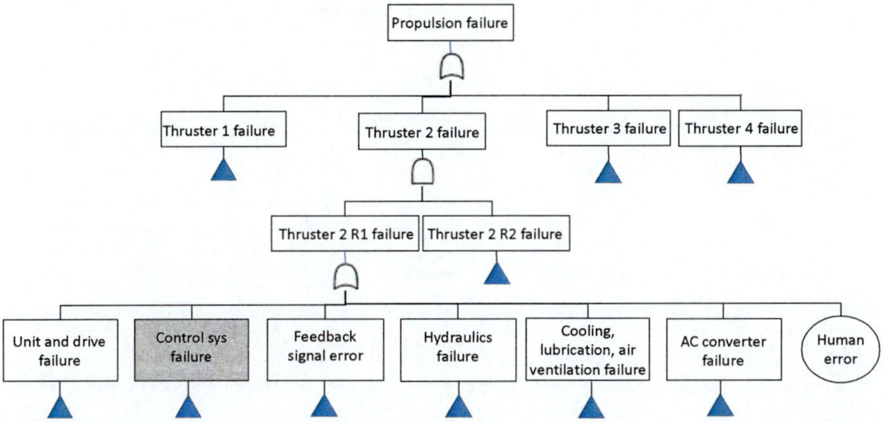

Fig. 9.4 Fault tree of a propulsion system in a DP drilling vessel, highlighted event has the highest failure frequency. (R1: redundant 1, R2: redundant 2, sys: system, AC: alternating current)

by DP capable lay barges or pipe laying ships. On the barge, the pipe is assembled or sometimes even constructed through a number of phases of welding which are carried out in a linear pipe manufacturing facility. Figures 9.10, 9.11 and 9.12 present propulsion, reference, and power system fault trees of DP cable/pipe laying vessels, respectively.

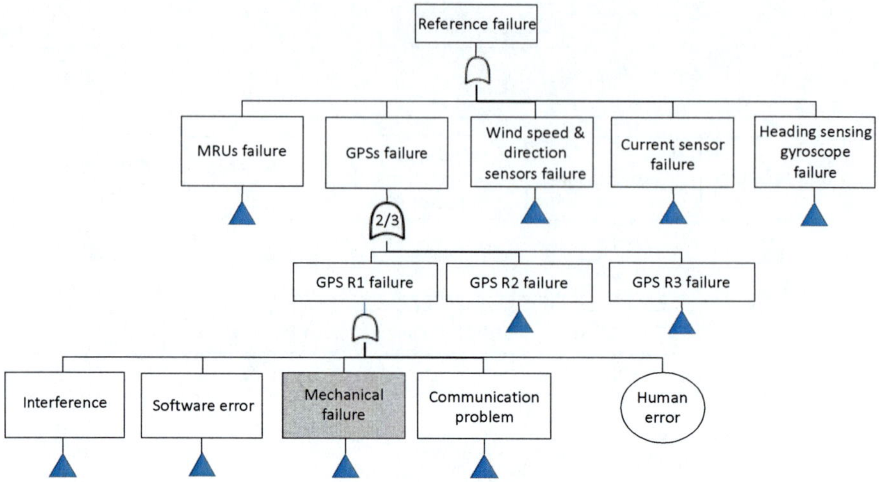

Fig. 9.5 Fault tree of a reference system in a DP drilling vessel, highlighted event has the highest failure frequency. (MRU: motion reference unit, GPS: global positioning system, Ri: Redundant i)

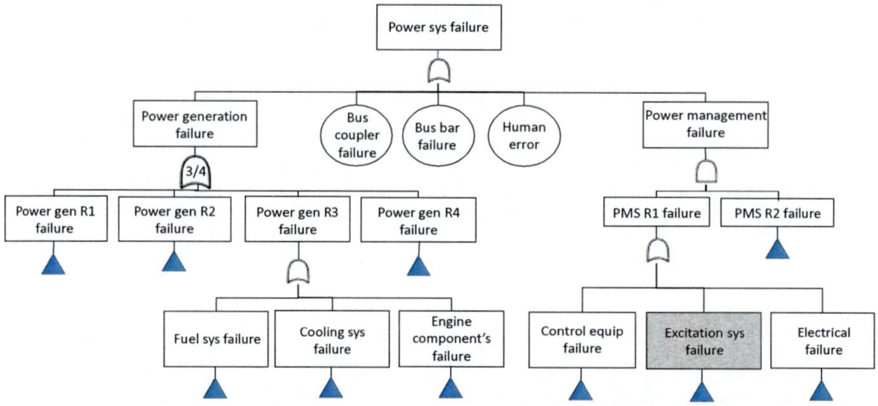

Fig. 9.6 Fault tree of a power system in a DP drilling vessel, highlighted event has the highest failure frequency. (Ri: redundant i, gen: generator, PMS: power management system, sys: system)

A shuttle tanker, as the name suggests, is a ship designed for transportation of oil or gas from an offshore field to wherever it may be required. It is highly sophisticated and well equipped with loading and offloading equipment compatible with the oil field infrastructures. The position keeping for longer durations of such ships is generally carried out on DP. Well backed-up safety systems ensure that the potentially flammable oil and natural gas are handled safely. Figures 9.13 and 9.14 show propulsion and power system fault trees of DP cable/pipe laying vessels, respectively.

In the design phase, these fault trees may provide a better understanding of potential faults of main systems and advance the technology or consider redundant systems

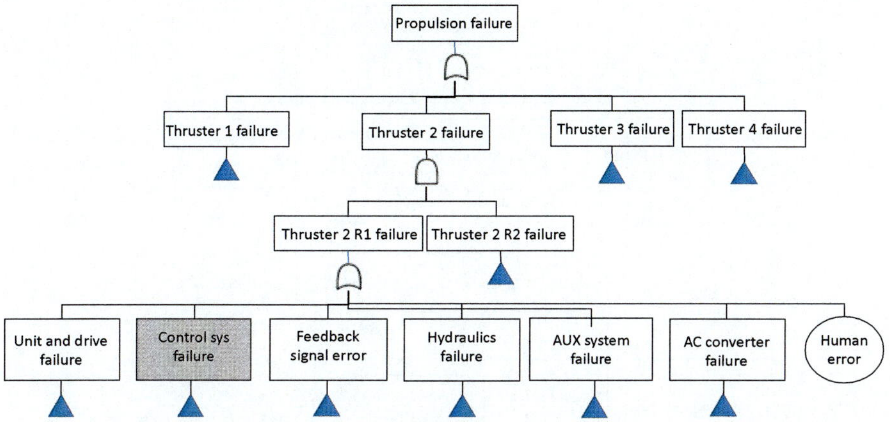

Fig. 9.7 Fault tree of a propulsion system in a DP diving vessel, highlighted event has the highest failure frequency. (R1: redundant 1, R2: redundant 2, sys: system, AC: alternating current)

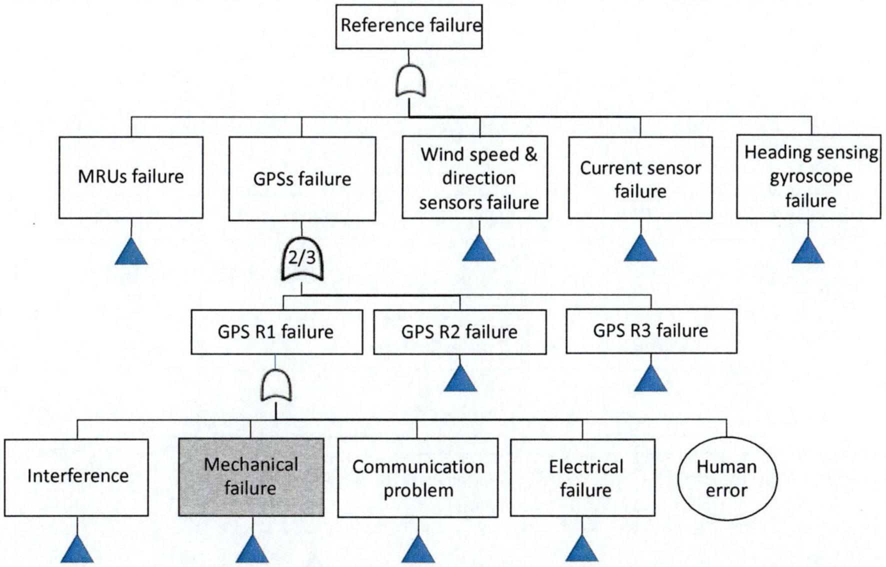

Fig. 9.8 Fault tree of a reference system in a DP diving vessel, highlighted event has the highest failure frequency. (MRU: motion reference unit, GPS: global positioning system, Ri: Redundant i)

in their design. Table 9.1 presents the distribution (percentage) of the main system faults and human errors among the studied incidents derived from international marine contractor's association (IMCA) reports between 2004 and 2015, with over

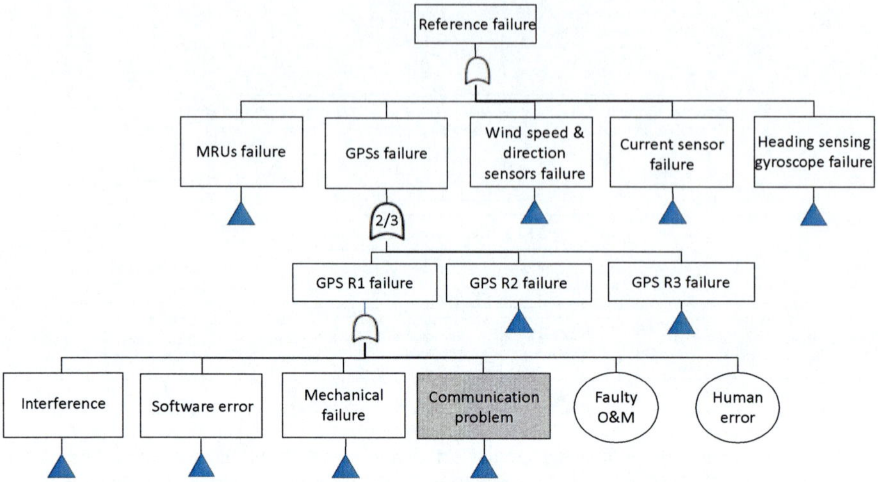

Fig. 9.9 Fault tree of a reference system in a DP ROV vessel, highlighted event has the highest failure frequency. (MRU: motion reference unit, GPS: global positioning system, Ri: Redundant i)

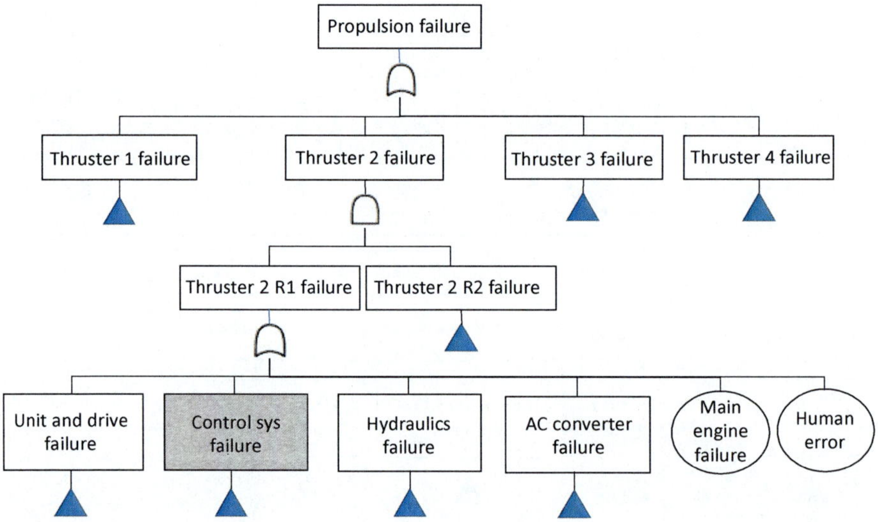

Fig. 9.10 Fault tree of a propulsion system in a DP cable/pipe laying system, highlighted event has the highest failure frequency. (R1: redundant 1, R2: redundant 2, sys: system, AC: alternating current)

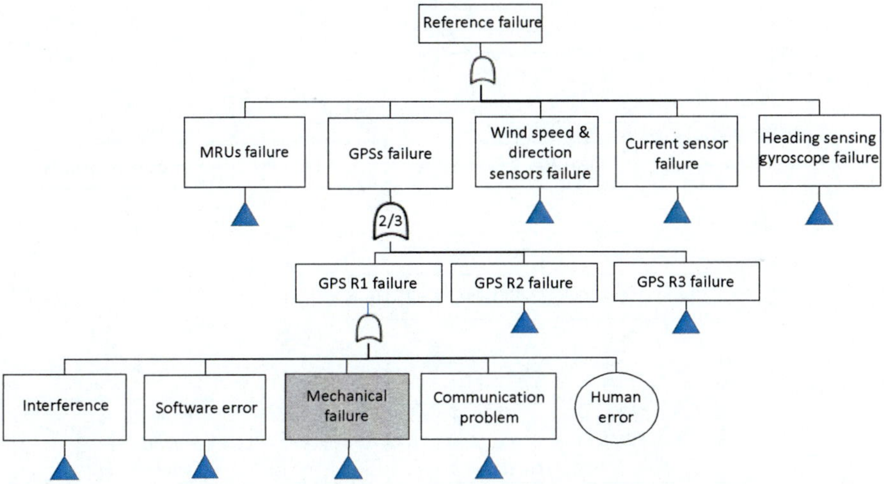

Fig. 9.11 Fault tree of a reference system in a DP cable/pipe laying system, highlighted event has the highest failure frequency. (MRU: motion reference unit, GPS: global positioning system, Ri: Redundant i)

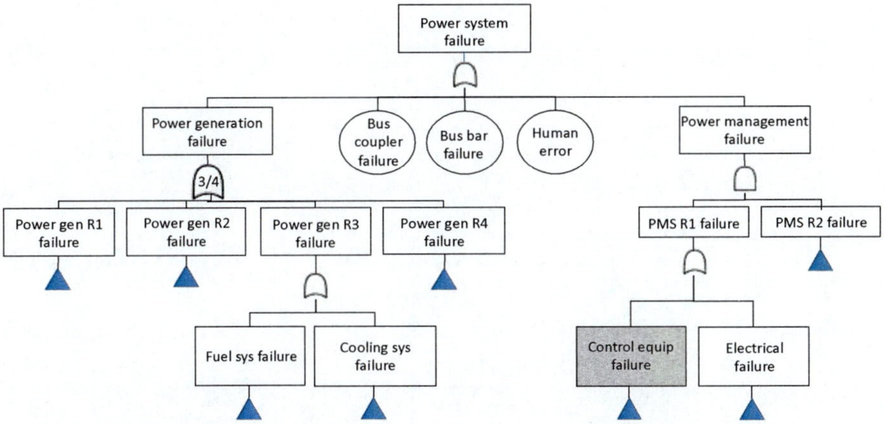

Fig. 9.12 Fault tree of a power system in a DP cable/pipe laying system, highlighted event has the highest failure frequency. (Ri: redundant i, gen: generator, PMS: power management system, sys: system)

700 incident reports.[1] The values in bold indicate the system with highest failure rate in different DP system types.

[1] The International Marine Contractors Association. (2005–2017). Dynamic Positioning Station Keeping Review, Incidents and events reported for (2004–2015). International Marine Contractors Assosiation.

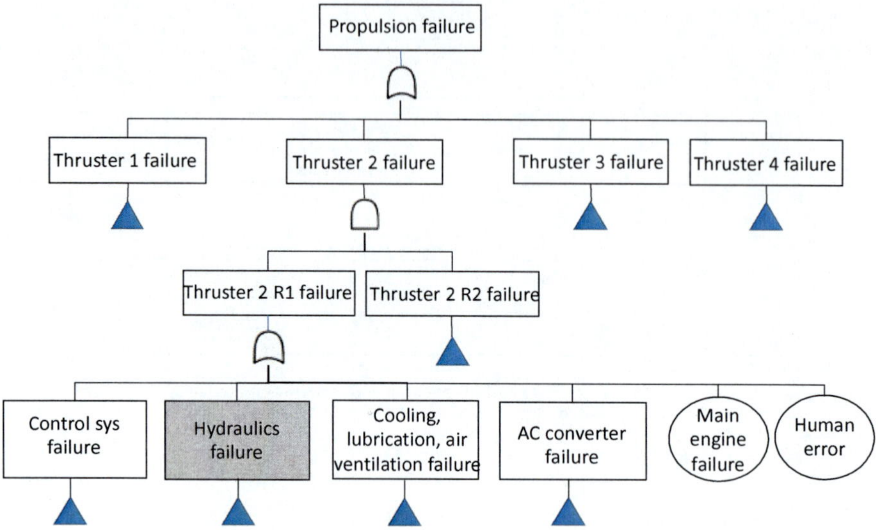

Fig. 9.13 Fault tree of a propulsion system in a DP shuttle tanker system, highlighted event has the highest failure frequency. (R1: redundant 1, R2: redundant 2, sys: system, AC: alternating current)

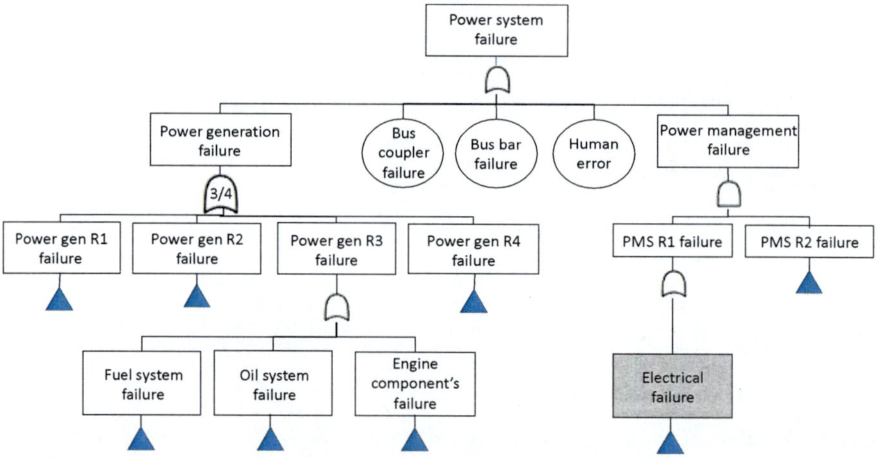

Fig. 9.14 Fault tree of a power system in a DP shuttle tanker system, highlighted event has the highest failure frequency. (Ri: redundant i, gen: generator, PMS: power management system, sys: system)

The vessel specific fault trees reflect the risk level of a particular DP type more accurately than the generic trees, as the failure frequencies of basic events are evaluated for that type of the DP system.

Table 9.1 System failure percentage in different dynamic positioning types[7]

Vessel type	Thruster (%) fault	Ref. faults (%)	Computer faults (%)	Power faults (%)	Human error
Shuttle tanker	9	**53**	15	13	8
Supply	**30**	27	11	17	10
Drilling	20	11	8	**26**	16
Diving	**32**	11	18	16	13
ROV (IMR)[a]	**28**	21	18	20	8
Cable/pipe laying	20	14	19	**24**	19

[a] ROV (IMR) refers to remotely operated vehicle (inspection, maintenance, and repair)

Glossary

Algorithm An algorithm is a step-by-step guide outlining how to do something [1].

Artificial intelligence (AI) It is the ability of a computer to simulate human thinking and understanding. AI can analyze and learn [1].

Complexity A causal chain with many intervening variables and feedback loops that do not allow the understanding or prediction of the system's behavior on the basis of each component's behavior [2].

Consequences The effects of the activity with respect to the values defined (such as human life and health, environment, and economic assets), covering the totality of states, events, barriers, and outcomes. The consequences are often seen in relation to some reference values (planned values, objectives, etc.), and the focus is often on negative, undesirable consequences [2].

Events The occurrence or change of a particular set of circumstances such as a system failure, an earthquake, an explosion, or an outbreak of a pandemic [2].

Failure Termination of the ability of an item to perform a required function [3].

Fault The inability of an entity to perform its required function, resulting in unavailability [4].

Machine learning Machine learning is a subsection of AI that focuses on allowing machines to learn new skills. So, over time, a computer becomes able to solve new problems and gets better at completing given tasks [1].

Model A model of an object (e.g., activity, system) is a simplified representation of this object. A probability model is a special type of models, based on frequentist probabilities (often referred to as chances in a Bayesian context) [2].

Model uncertainty Uncertainty about the model error, i.e., about the difference between the model output and the true value being modeled [2].

Offshore drilling vessels A drilling vessel will use DP to remain in a fixed location while drilling in deep water [3].

Probability A measure for representing or expressing uncertainty, variation, or beliefs, following the rules of probability calculus [2].

T. Parhizkar et al., *Online Probabilistic Risk Assessment of Complex Marine Systems*, Springer Series in Reliability Engineering,
https://doi.org/10.1007/978-3-030-88098-9

Risk We consider a future activity [interpreted in a wide sense to also cover, for example, natural phenomena], for example, the operation of a system, and define risk in relation to the consequences (effects, implications) of this activity with respect to something that human's value. The consequences are often seen in relation to some reference values (planned values, objectives, etc.), and the focus is often on negative, undesirable consequences. There is always at least one outcome that is considered as negative or undesirable [2].

Risk analysis Systematic process to comprehend the nature of risk and to express the risk, with the available knowledge. Risk analysis is often also understood in a broader way, in particular in the Society for Risk Analysis (SRA) community: Risk analysis is defined to include risk assessment, risk characterization, risk communication, risk management, and policy relating to risk, in the context of risks of concern to individuals, to public and private sector organizations, and to society at a local, regional, national, or global level [2].

Risk assessment Systematic process to comprehend the nature of risk, express, and evaluate risk, with the available knowledge [2].

Risk evaluation Process of comparing the result of risk analysis against risk criteria to determine the significance and acceptability of the risk [2].

Risk-influencing factor Risk-influencing factors (RIFs) are background factors that influence the causes and/or the development of an accident [5].

Risk management Activities to handle risk such as prevention, mitigation, adaptation, or sharing. It often includes trade-offs between costs and benefits of risk reduction and choice of a level of tolerable risk [2].

Risk model An informative representation of a system or activity that provides the system risk level/spectrum.

Rule-based system Arule-based system is one that applies human-made rules to run the desired action [1].

Safety The antonym of risk (the safety level is linked to the risk level; a high safety means a low risk and vice versa) [2].

Safety analysis Systematic process to comprehend the nature of the safety of a system and to express the safety level. Systematic process to determine the degree of risk reduction that is sufficient to obtain a "safe system" [2].

Security The antonym of risk when restricting the concept of risk to intentional acts by intelligent actors (the security level is linked to the risk level; a high security level means a low risk and vice versa) [2].

Uncertainty Imperfect or incomplete information/knowledge about a hypothesis, a quantity, or the occurrence of an event [2].

References

1. An automation glossary, https://www.thinkautomation.com/eli5/an-automation-glossary/. Date Accessed: June 16, 2021
2. Aven T, Ben-HaimY, Boje Andersen H, Cox T, Droguett EL, GreenbergMet al (2018) Society for risk analysis glossary. Society for risk analysis
3. ISO 14224:2016, Petroleum, petrochemical and natural gas industries—Collection and exchange of reliability and maintenance data for equipment
4. Smith DJ (2017) Reliability, maintainability and risk: practical methods for engineers. Butterworth-Heinemann, Oxford, UK
5. Rausand, Haugen. (2020). Risk assessment theory, practice and assessment. 2nd edition. Wiley, NJ, USA

T. Parhizkar et al., *Online Probabilistic Risk Assessment of Complex Marine Systems*, Springer Series in Reliability Engineering,
https://doi.org/10.1007/978-3-030-88098-9

Index

T. Parhizkar et al., *Online Probabilistic Risk Assessment of Complex Marine Systems*, Springer Series in Reliability Engineering,
https://doi.org/10.1007/978-3-030-88098-9